Industrial Instruments for Measurement and Control

Fred Powell

Industrial Instruments for Measurement and Control

SECOND EDITION

by

Thomas J. Rhodes

Formerly Assistant Director, Research and Development Division
Uniroyal, Inc.

Revised by Grady C. Carroll

Senior Instrument Engineer
Ford, Bacon & Davis Construction Corp.

McGRAW-HILL BOOK COMPANY

New York St. Louis San Francisco Düsseldorf Johannesburg
Kuala Lumpur London Mexico Montreal New Delhi Panama
Paris São Paulo Singapore Sydney Tokyo Toronto

Library of Congress Cataloging in Publication Data

Rhodes, Thomas J
Industrial instruments for measurement and control.

1. Engineering instruments. 2. Chemical engineering—Apparatus and supplies. 3. Automatic control. I. Carroll, Grady C. II. Title
TA165.R45 1972 620'.0044 71-172663
ISBN 0-07-052121-2

4 5 6 7 8 9 0 KPKP 7 9 8

The editors for this book were Tyler G. Hicks and Lydia Maiorca, the designer was Naomi Auerbach, and its production was supervised by George E. Oechsner. It was set in Caledonia by The Maple Press Company.

It was printed and bound by The Kingsport Press.

To my devoted wife Grace

Contents

Preface

Industrial Instruments for Measurement and Control has been revised, with deletions and additions, to cover the more modern process-measuring-and-controlling equipment that was not available when Mr. Thomas J. Rhodes prepared the first edition in 1941. However, many of Rhodes' original work has been retained because much of the analyses of control problems are just as practicable today as they were in 1941.

The text has been prepared for use by technical-vocational schools as a textbook for those who are preparing to enter the vast field of industrial instrumentation, as well as for the industrial instrument application engineer. It covers information that can be used in organizing instrument maintenance departments, organization of instrument departments, practical functional organization charts, and preventive maintenance schedules. The industrial instrument engineer will find techniques of analyzing process control problems and selecting the proper measuring and controlling equipment to design into a control system.

Grady C. Carroll

Acknowledgments

THE FOLLOWING COMPANIES have contributed material for this book, and their assistance is gratefully acknowledged.

Alnor Instrument Company
Chicago, Illinois

American Meter Company
Erie, Pennsylvania

Bailey Meter Company
Wickliffe, Ohio

Beckman Instrument Company
Fullerton, California

BIF, Unit of General Signal Corporation
Providence, Rhode Island

Black, Sivalls & Bryson, Incorporated
Tulsa, Oklahoma

The Bristol Company, Division of American Chain & Cable Company, Inc.
Waterbury, Connecticut

Brooks Instrument Company, Inc.
Hatfield, Pennsylvania

Chandler Engineering Company
Tulsa, Oklahoma

Daniel Industries, Incorporated
Houston, Texas

Fischer & Porter Company
Warminster, Pennsylvania

Fischer Governor Company
Marshalltown, Iowa

The Foxboro Company
Foxboro, Massachusetts

Hagan/Computer Systems, Division of Westinghouse Electric Corporation
Pittsburg, Pennsylvania

Honeywell, Incorporated
Fort Washington, Pennsylvania

Industrial Instrument Corporation
Austin, Texas

In-Val-Co, Division of Combustion Engineering, Incorporated
Tulsa, Oklahoma

Leeds & Northrup Company
North Wales, Pennsylvania

Manning, Maxwell & Moore, Incorporated
Stratford, Connecticut

Moore Products Co.
Spring House, Pennsylvania

The Ohmart Corporation
Cincinnati, Ohio

Potter Aeronautical Corporation
Union, New Jersey

Robertshaw Controls Company
Anaheim, California

Rockwell Manufacturing Company
Pittsburg, Pennsylvania

Taylor Instrument Companies
Rochester, New York

The Vapor Recovery Systems Company
Compton, California

Industrial Instruments for Measurement and Control

GENERAL INTRODUCTION

The material presented in this book consists largely of information on instruments used for the measurement and control of the four basic physical and thermal quantities that are the concern of the chemical, mechanical, instrument, or electrical engineer, or of anyone who is involved in the design, operation, and maintenance of industrial processing plants, such as petrochemical, steam generation, power generation, fertilizer, sulphuric acid, and nitrous acid, etc. These quantities are: (1) temperature, (2) pressure, (3) flow, and (4) liquid level.

Many additional quantities are measured and controlled, and some of these are covered in this book; but others (such as weight) involve equipment not classed as "instruments" in the somewhat restricted definition of the word as used here. However, by far the largest number of industrial instruments are confined to the measurement and control of the four quantities listed above, and if this group of instruments is thoroughly mastered in theory and application, most other industrial instruments can be easily understood and applied in practice. For this reason several of the 13 chapters of this book are devoted to these basic instruments, and a chapter is devoted to a short description of each of several special measuring and controlling instruments.

CHAPTER ONE

Temperature-measuring Instruments

THE MEASUREMENT OF TEMPERATURE is one of the most important functions of an industrial instrument. Temperature standards are available for maintaining the accuracy of temperature-measuring instruments, such as thermometers and pyrometers. Furthermore, since a temperature standard must be referred to a basic temperature scale, it is logical to discuss briefly the theoretical basis upon which the temperature scale is built.

The temperature scale that satisfies the most rigorous logic is the thermodynamic scale, defined by Lord Kelvin as that scale wherein the absolute values of two temperatures are to each other in the proportion of the heat taken in to the heat rejected in a reversible thermodynamic engine working with a heat source and a refrigerator at the higher and lower of the temperatures, respectively. The size of the unit degree is not established in this definition, but it is usually arbitrarily set by the use of the interval from the freezing point to the boiling point of pure water as equivalent to 100°C. This ideal scale is never achieved exactly in practice by any temperature-measuring instrument, and there are few thermometers or pyrometers that use the Kelvin principle in their operation. However, standard thermometers have been built that deviate only a small amount from the true thermodynamic scale, and the corrections that must be applied are known with considerable accuracy. The thermodynamic scale may therefore be considered as the basic temperature scale upon which all temperature standards are built.

Fundamental Working Parts That Comprise All Pressure-actuated Thermometers

All pressure-actuated thermometers are made up of five distinct related parts which serve to translate a change in temperature in the medium being measured into a continuous record of that change on a time-temperature chart. These five parts are as follows:

1. A sensitive pressure bulb to be immersed in the medium to be measured and containing the actuating pressure medium
2. A capillary tube connecting the bulb to the recorder
3. An actuating mechanism in the recorder to respond to changes in pressure
4. A linkage to multiply the movement of the actuating mechanism and draw a line on the recording chart
5. A chart mechanism to hold a recording chart and revolve it at a fixed speed

The Thermometer Bulb

Changes in the pressure of the medium filling the thermometer system take place because of heat absorbed or removed from the sensitive bulb. This increase or decrease in pressure is then transmitted through the capillary tubing to the instrument, where a change of temperature is recorded. The paramount consideration in the design of this part is to enable the bulb to respond as quickly as possible to any difference in temperature between the bulb and the medium being measured. Performance of the measuring system is exactly the same as that of a pressure-measuring system: the pressure in the filled system is proportional to the temperature applied to the temperature-sensing bulb.

Bimetallic Strip-actuating Element

The bimetallic strip shown in Fig. 1-1 is one type of actuating mechanism sometimes used for recording thermometers which does not operate on changes in pressure, resistance, or millivolts. This mechanism is used to record only temperatures at or near atmospheric pressure and only where the recorder can be placed in the gas to be measured.

The bimetallic, or thermostatic, strip is constructed of two strips of different metals having a large difference in coefficient of lineal expansion and welded together along their entire length. Any change in temperature produces a greater expansion in the strip with the higher coefficient;

since one side of this strip is bound to the low-expansion strip, it is restrained on this side and tends to warp. Conversely, the strip of low coefficient is stretched on the side in contact with the high-coefficient strip.

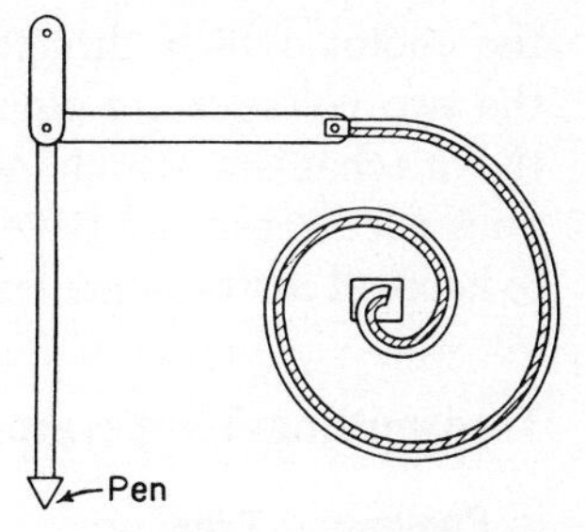

Fig. 1-1 Bimetallic strip for recording thermometer.

Requirements for Liquids of the Liquid-filled-type Thermometer

Requirements for a desirable liquid to be used in recording and indicating thermometers as a filling medium are as follows:

1. The vapor pressure must be negligible over the temperature range for which it is to be used.
2. The coefficient of cubical expansion should be high.
3. The liquid must be chemically inactive with respect to the metal in the thermometer system.
4. The liquid should have a low specific gravity, a low specific heat, and a high coefficient of heat conductivity.
5. The liquid must be incompressible.

The first three of these requirements are the most important, and any liquid with good characteristics in these categories is preferred regardless of its specific gravity and specific heat.

Modern Self-balancing Potentiometer Recorders

Years ago, instrument engineers depended upon the galvanometer to detect unbalance between the known and unknown voltage and through mechanical means, to rotate a slide-wire (or some other component) to balance the known and unknown voltages so that the galvanometer would always be at zero when the instrument was measuring exactly the millivolts coming from a thermocouple. However, developments in the past years have almost eliminated the galvanometer for detecting unbalance between the known and unknown voltages. The dc voltage coming from the thermocouple is passed through a chopper or converter and is converted to a square-wave ac voltage. From there it goes into an input transformer and from the transformer into a voltage-amplifying system consisting of tubes or transistors; the square wave is rounded off so that it is not too far from a 60-cycle sine wave. This is then passed through a phase detector so that it can be determined whether the thermocouple voltage is higher or lower than the slide-wire voltage. From there it goes into another part of the circuit that consists of power amplification, and it comes out as a voltage amplified to such a point that it will drive a small two-phase motor. The phase detector tells the motor which direction to turn so that for all practical purposes the two voltages are equal in value. The manufacturers have simplified the mechanism which now measures temperature, and the instrument does not necessarily have to be level; however, it is always good practice to keep all instruments level.

Transmitting Temperature Instruments

Pneumatic Type

Bailey Meter Company

PRINCIPLE OF DESIGN: Figure 1-2 is a front view of Bailey Meter Company's KT13 pneumatic temperature transmitter, a motion-balance type which employs a nozzle, baffle, or vane, and a follow-up or restoring bellows to convert a measured temperature into an output air pressure

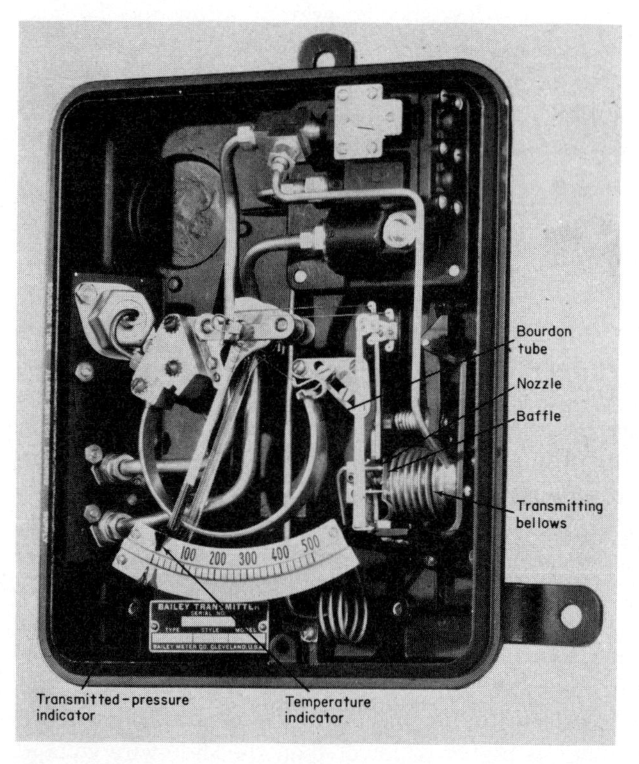

Fig. 1-2 Front view of Bailey Meter Company type KT13 pneumatic temperature transmitter. (*Bailey Meter Company*)

of 3 to 27 or 3 to 15 psi, depending upon which range is more desirable. The temperature-measuring system consists of a Bourdon tube, capillary, and bulb which is filled with helium and sealed.

The model KT13 temperature transmitter incorporates one item not usually found in indicating transmitters: the pneumatically transmitted signal indicator along with the temperature indicator. This combination of indicators is convenient for checking the accuracy of the transmitted signal because whenever the two pointers do not coincide, an error in the transmitted signal is indicated (which can be detected by observing the position of the pointers).

The Bristol Company, Division of American Chain & Cable Company, Inc.

PRINCIPLE OF DESIGN: Figure 1-3 is a model 9E7 pickup unit of the Bristol radiation pyrometer. The Bristol Company has categorized its pickup heads as a low-temperature head, model 9E7 (shown in Fig. 1-3), which has a range of 0 to 1100°F, and a high-temperature head, model 9E9 (shown in Fig. 1-4), which has a range of 0 to 3600°F.

Both are noncontact-type measuring devices and work on the principle that the radiated energy emitted by a hot body is in definite relationship to its temperature. The total radiated energy of an object is measured and converted to a millivoltage signal which can then be displayed on any Bristol Dynamaster* recorder in units of temperature.

* Registered trade name, The Bristol Company.

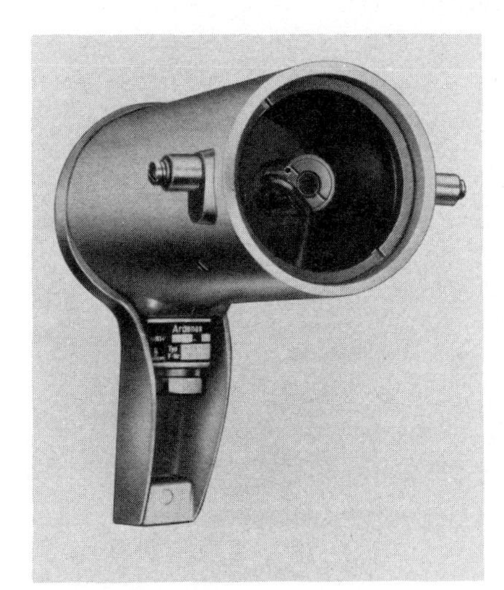

Fig. 1-3 Radiation pickup head for temperature from 0 to 1100°F. (*The Bristol Company, Division of American Chain & Cable Company, Inc.*)

Fig. 1-4 Radiation pickup head for temperatures from 0 to 3600°F. (*The Bristol Company, Division of American Chain & Cable Company, Inc.*)

The Foxboro Company*

PRINCIPLE OF DESIGN: Figure 1-5 is a Foxboro Company emf Dynalog† recorder which is used in conjunction with a thermocouple to measure and record temperatures. The Dynalog functions without the use of a slide-wire and dry-cell battery; instead an air condenser and a power unit drive, which Foxboro terms a Dynapoise,‡ are used. As seen in the schematic wiring diagrams in Figs. 1-6 and 1-7, the Dynalog circuit is a modified bridge circuit with condensers forming two of its legs.

* Grady C. Carroll, *Industrial Instrument Servicing Handbook,* 1st ed., McGraw-Hill Book Company, New York, 1960, pp. 5-44, 5-45.

† Registered trade name, the Foxboro Company.

‡ Registered trade name, the Foxboro Company.

Fig. 1-5 Temperature recorder. (*The Foxboro Company*)

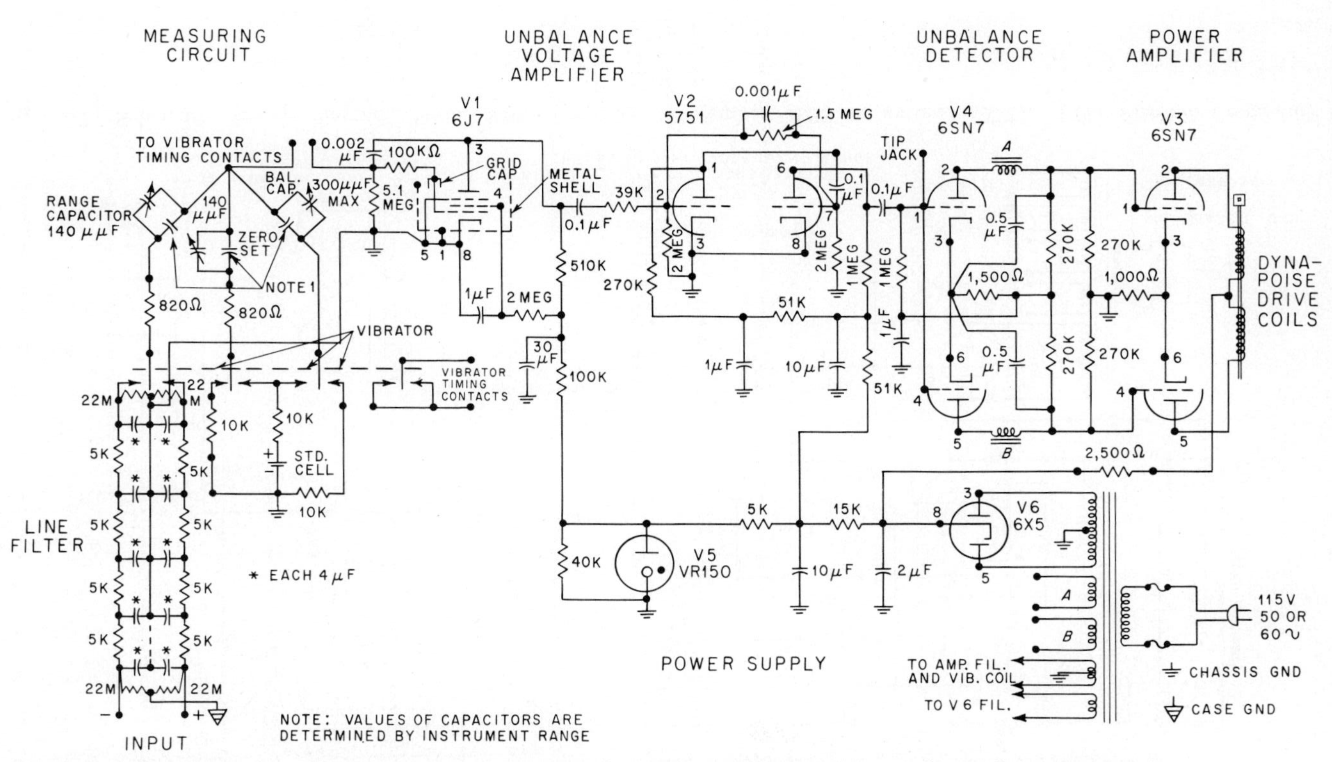

Fig. 1-6 Schematic wiring diagram, Series 9300 Recorder (since autumn 1957). (*The Foxboro Company*)

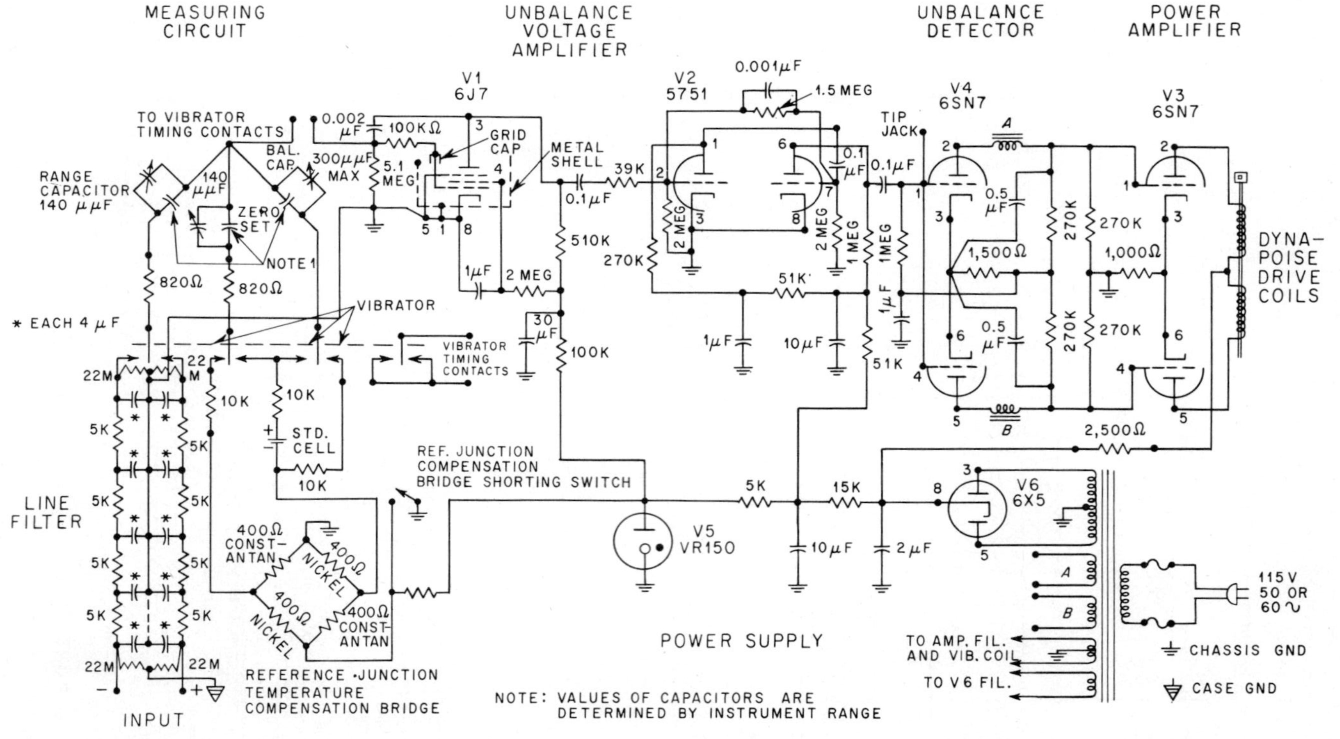

Fig. 1-7 Schematic wiring diagram, Series 9300 Recorder (December 1955 to autumn 1957). (*The Foxboro Company*)

The condenser in the left side of the bridge is manually adjusted to give the instrument the desired range span. The balancing condenser in the right leg is positioned by the Dynapoise unit shown in Fig. 1-8. The condenser in the center tap of the bridge is manually adjusted to produce a zero reading of the instrument when a zero emf is applied to the input connections indicated on the schematic.

A standard cell has its output connected to three contacts which are a part of a synchronous converter or vibrator driven by the 115-volt 60-cycle power supply. During one-half of a cycle the standard cell is connected across the balancing condenser in the right leg of the bridge, and one-half of the emf source is connected across the range condenser in the left leg of the bridge. The zero capacitor is short-circuited through the grid resistor of the 6J7 tube. During the next half cycle, the leg containing the balancing condenser is short-circuited through the grid resistor of the 6J7 tube. One-half of the emf source is connected across the capacitor with reversed polarity, and the standard cell is connected across the zero condenser. The direction of current flow change through the grid resistor of the 6J7 tube determines whether its grid is more or less negative in respect to the cathode of the tube, which in turn determines the amount of plate current passed by the tube. The direction of current change through the 6J7 grid resistor also determines whether the output of the tube is exactly in phase or exactly out of phase with the 60-cycle power driving the vibrator.

THE FOXBORO COMPANY*

PRINCIPLE OF DESIGN: Figure 1-9 is a Foxboro model 12A nonindicating temperature transmitter of the pneumatic force-balance type which

* Grady C. Carroll, *Industrial Process Measuring Instruments,* 1st ed., McGraw-Hill Book Company, New York, 1962, pp. 256–257.

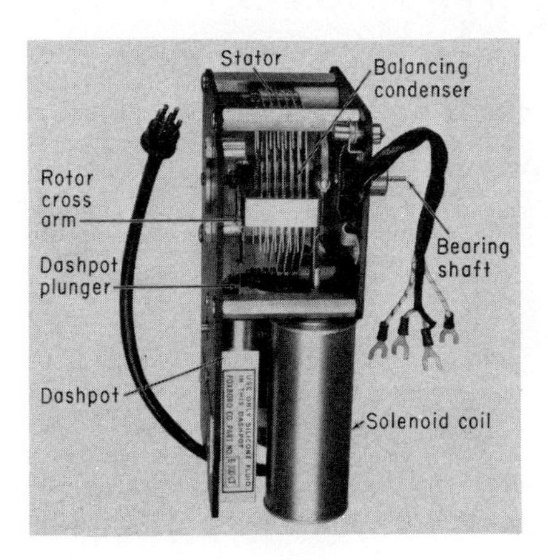

Fig. 1-8 Dynapoise unit. (*The Foxboro Company*)

transmits a pneumatic signal in a 3- to 15-psi range, the value of which is proportional to the measured temperature. A gas-filled bulb, capillary, and bellows element constitute the temperature-responsive system.

In operation, the following steps occur simultaneously, but are listed separately for the purpose of explanation. If a rise in the measured temperature is assumed:

- The pressure increases in the thermal element capsule, thereby increasing the force on the lower end of the force bar.
- The force bar thus tends to rotate about the flexure fulcrum so that its opposite end approaches the nozzle and increases the back pressure therein.
- The back pressure from the nozzle is amplified through a conventional relay and then applied to the feedback bellows. This pressure is also the transmitter output.
- The resulting pressure increase in the feedback bellows produces a restoring force on the force bar which opposes that of the measuring element. The system is thus rebalanced at a new output pressure,

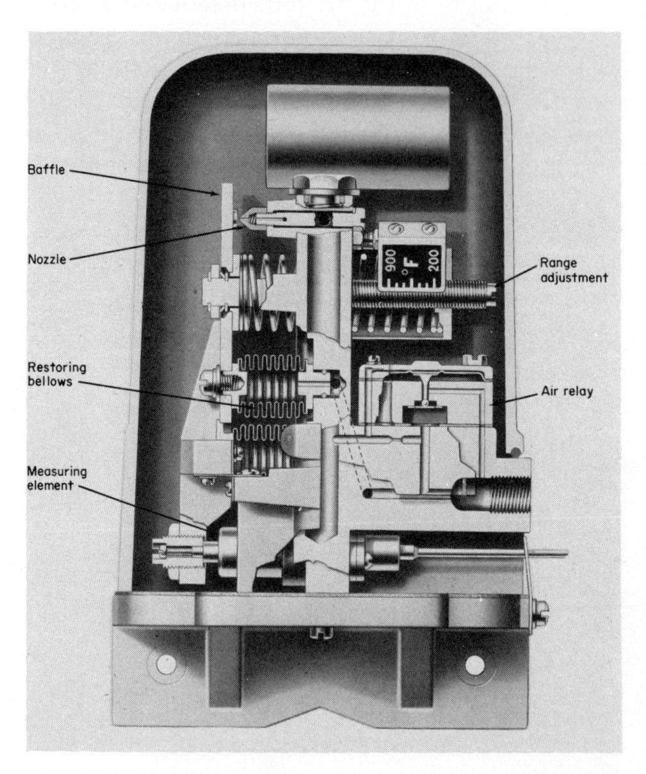

Fig. 1-9 Foxboro model 12A pneumatic temperature transmitter. (*The Foxboro Company*)

which has changed an amount proportional to the change in the measured temperature.

There are two additional forces which act on the force bar. One is that applied by the span-elevation spring; the other is produced by the ambient-temperature- and the barometric-pressure-compensating bellows. The effect of the span-elevation spring is to permit elevation and depression of the base point, which is the temperature at which the transmitter output pressure is 3 psi. The purpose of the ambient-temperature- and barometric-pressure-compensating bellows is to produce an opposing force when the ambient temperature or pressure is felt by the element capsule.

Moore Products Co.

PRINCIPLE OF DESIGN: Figure 1-10 is a Moore Products Co. Nullmatic* temperature transmitter which consists of a gas-filled thermal system operating in conjunction with a balancing bellows and air supply to transmit an air pressure having direct linear relation to the bulb temperature. The transmitting system uses the "null-balance" principle, which

* Registered trade name, Moore Products Company.

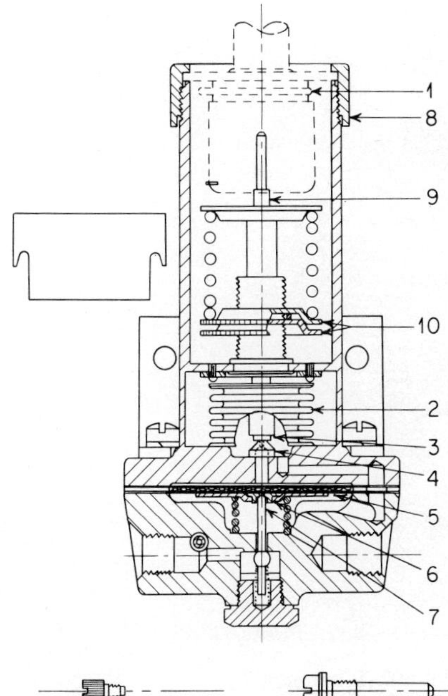

Fig. 1-10 Moore Products Co. pneumatic transmitter. (*Moore Products Co.*)

avoids the effect of bellows nonlinearity by maintaining the bellows in a nearly fixed position for any value of measured temperature.

Taylor Instrument Companies*

principle of design: Figure 1-11 shows a Taylor Instrument Companies model 202-T temperature transmitter of the pneumatic force-balance type. The instrument is a simple, rugged temperature transmitter which converts temperature, measured by a thermal-filled system, into a proportional air output pressure of 3 to 15 psi. The nonindicating transmitter then transmits this pressure to a receiving recorder or indicator and/or controller. A simple force beam eliminates compressure springs or balance weights.

The actuating element of the transmitter is the Bourdon tube of a mercury-filled system, shown schematically in Fig. 1-12. The Bourdon tube moves the force beam to reposition the baffle with respect to the nozzle. The nozzle back pressure is then amplified by a standard Taylor reversing relay and becomes the transmitter output. This output pressure also goes to the feedback bellows and acts as the restoring force, tending to bring the nozzle and baffle back to their original positions.

In applications where thermal or transmission lags are excessive, an adjustable derivative unit can be installed in the feedback line between

* *Ibid.*, pp. 262–264.

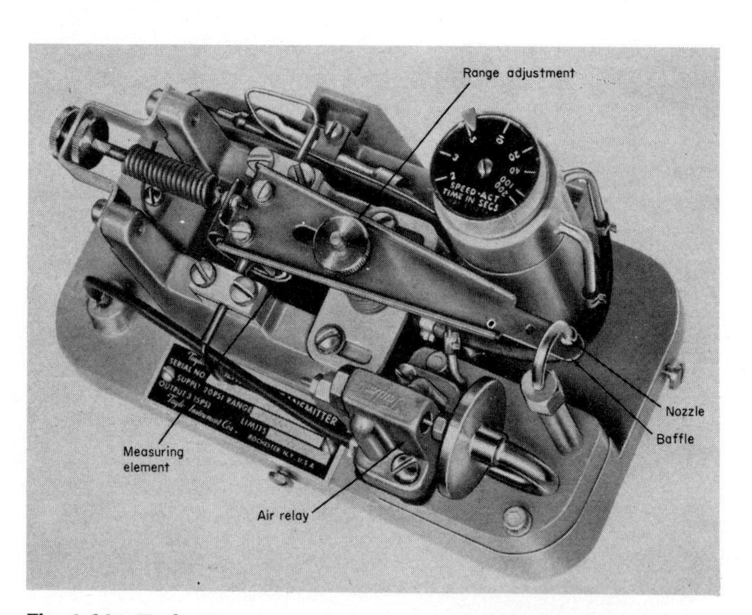

Fig. 1-11 Taylor Instrument Companies model 202-T temperature transmitter. (*Taylor Instrument Companies*)

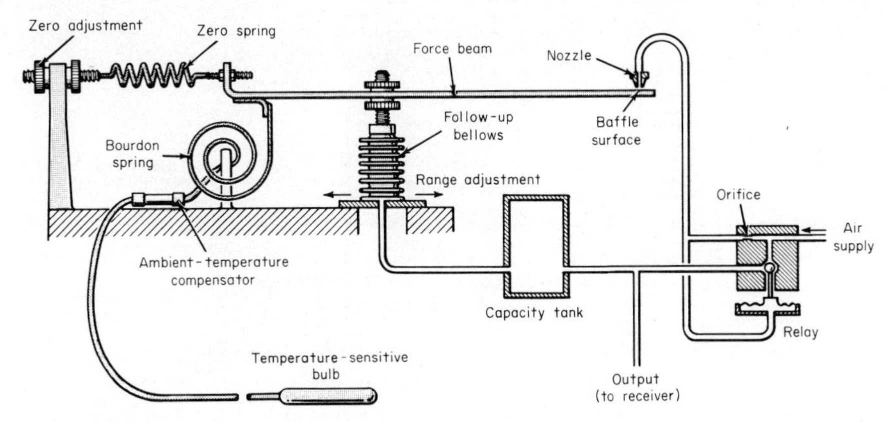

Fig. 1-12 Schematic of a Taylor temperature transmitter. (*Taylor Instrument Companies*)

the air relay and the follow-up bellows. With the derivative unit in place, a delay in feedback to the follow-up bellows can be created, which results in an exaggerated change in output pressure as long as the temperature is changing. Duration of the exaggerated output pressure is dependent upon the setting of the derivative adjustment. When the temperature becomes stable, the output pressure is again proportional to the temperature of the thermal-sensitive bulb.

Taylor Instrument Companies

PRINCIPLE OF DESIGN: Figure 1-13 is a cross-sectional view of a temperature transmitter designed by Taylor Instrument Companies. The Taylor Transaire* temperature transmitter operates on the force-balance principle, and the motion of all parts is imperceptible to the eye. The greatest movement is of the baffle tip, which is approximately 0.001 in. With a 50:1 amplification between the baffle and the capsular diaphragm, the total movement of the diaphragm is twenty-millionths of an inch for a full-range transmitter output.

The thermal system of the transmitter is filled with a monatomic gas. With an increase in temperature at the bulb of the thermal element, the gas in the system expands, and the pressure change is transmitted through the capillary to the capsular chamber. The upward motion exerted by this pressure change is transmitted to the spring plate and simultaneously to the force and motion levers by connecting wires.

An upward force on the motion lever produces a counterclockwise motion of the baffle to draw it away from the nozzle and at the same time reduces the back pressure in the nozzle line. The decrease in

* Registered trade name, Taylor Instrument Companies.

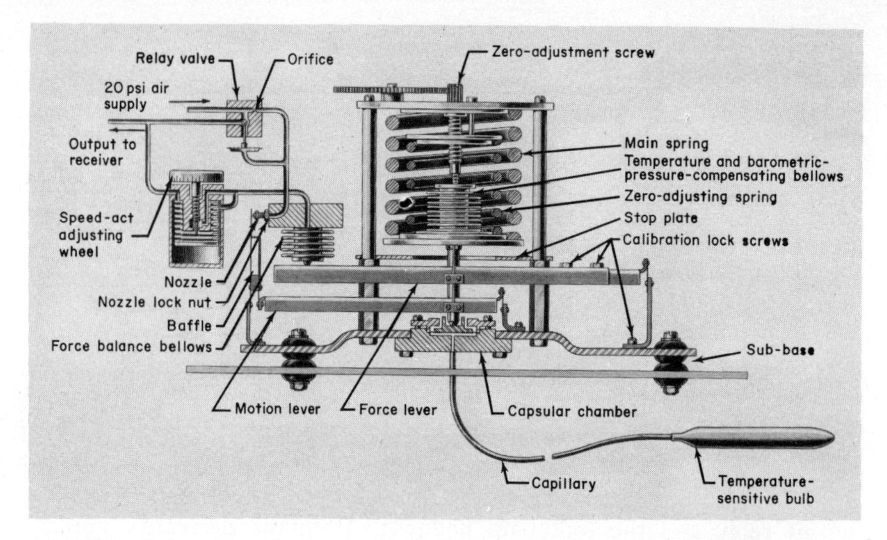

Fig. 1-13 Cross-sectional view of Taylor Instrument Companies pneumatic temperature transmitter. (*Taylor Instrument Companies*)

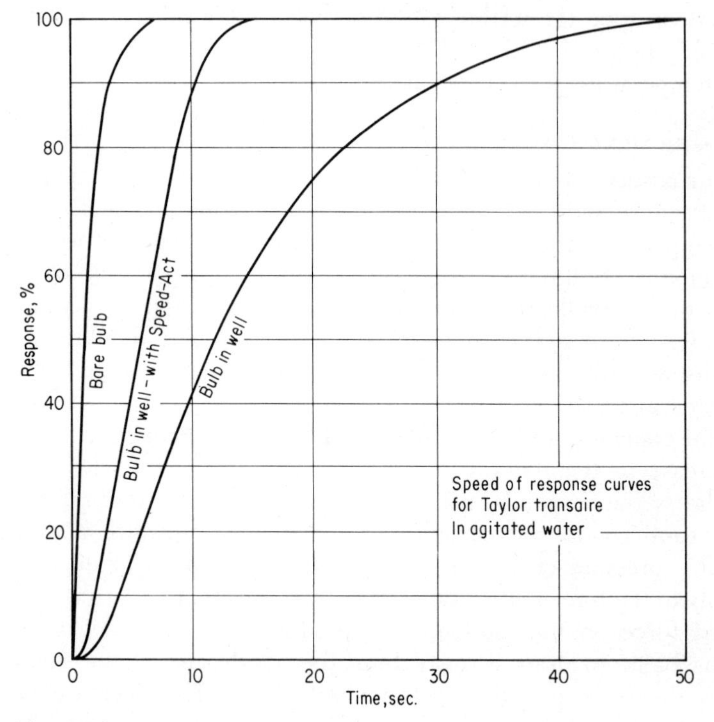

Fig. 1-14 Temperature response curve. (*Taylor Instrument Companies*)

back pressure produces an increased output pressure from the relay, which is transmitted to the balancing bellows and to the receiving instrument. An increased downward thrust, resulting from the pressure increase in the force-balance bellows, moves the force lever downward. At the same time, the spring plate is pulled downward by the connecting wires, which results in the restoration of the diaphragm to approximately its original position. Equilibrium is established when the thrust produced by the bellows is proportional to the thrust exerted by the thermal system. The pneumatic output pressure is, therefore, proportional to the measured temperature.

A Speed-Act* (derivative) unit can be inserted in the feedback line between the output and force-balance bellows to compensate the output for the time lag in the thermal system, which is illustrated by the curves in Fig. 1-14. When the Speed-Act unit is adjusted, a delay in the feedback pressure to the force-balance bellows is created, which makes the output signal more nearly follow the true temperature at the bulb than would be possible without the Speed-Act or derivative effect.

Taylor Instrument Companies†

PRINCIPLE OF DESIGN: A Taylor Instrument Companies model 700-T potentiometer transmitter is shown in Fig. 1-15. The instrument is de-

* Registered trade name, Taylor Instrument Companies.
† Carroll, *Industrial Process Measuring Instruments,* pp. 277–284.

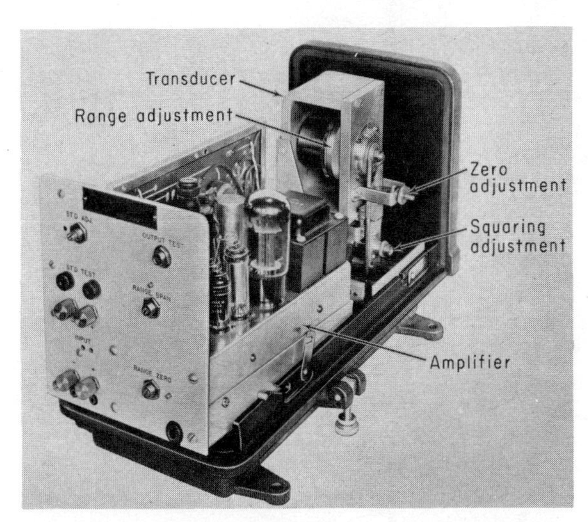

Fig. 1-15 Taylor Instrument Companies model 700-T potentiometer transmitter. (*Taylor Instrument Companies*)

signed to convert an electrical measurement made with such primary elements as resistance thermometer, thermocouple, and strain gauge into a pneumatic output signal of 3 to 15 psi. An electrical output in the form of milliamperes is also available for transmission or logging purposes.

The model 700-T transmitter is comprised of two basic units: an electronic potentiometer amplifier and an electropneumatic transducer.

Incorporated in the potentiometer is an amplifier which develops a direct output current from 0 to 5 ma, which is proportional to the output from a resistance element or a source of dc emf. This output can be transmitted to data-processing or other electrically actuated equipment.

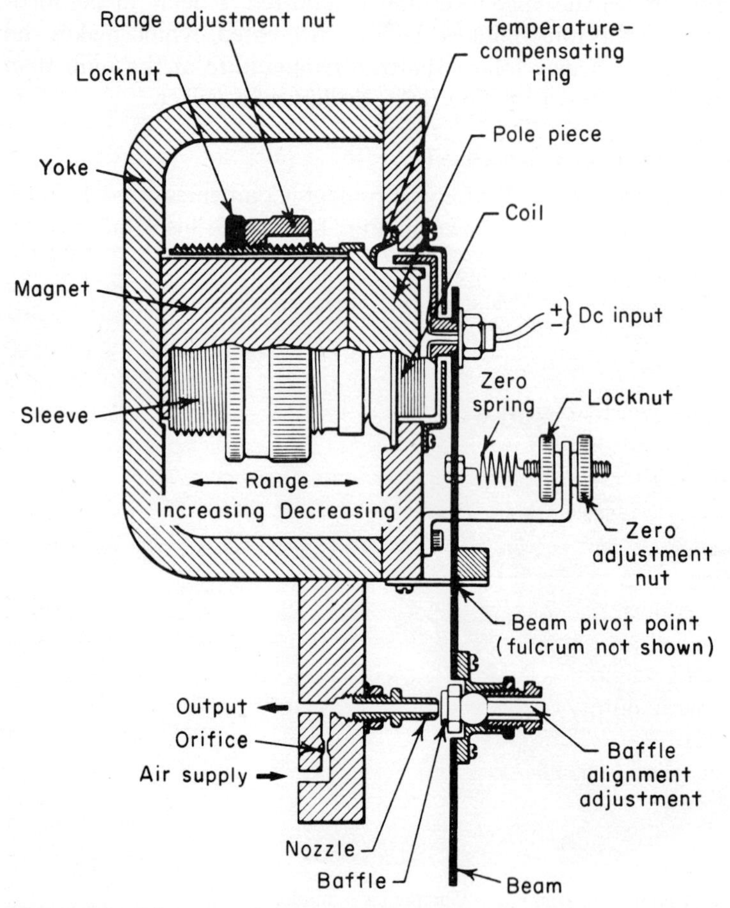

Fig. 1-16 Schematic diagram electropneumatic transducer. (*Taylor Instrument Companies*)

The integral electropneumatic transducer can be actuated simultaneously by the same output signal.

The transducer, which is available separately if desired, converts direct current in the range from 0 to 5 ma to a proportional air pressure in the range from 3 to 15 psi. This output air signal can then be used with pneumatic measuring and controlling instruments.

The potentiometer input circuit operates on the basic principles applying to the well-known Wheatstone-bridge circuit. This potentiometer uses the output current of the instrument as a feedback signal to the bridge circuit for effective balancing. Thus motorized slide-wires and their associated moving parts are eliminated.

The electropneumatic transducer incorporated in the model 700-T potentiometer transmitter converts the direct output current from the amplifier to a proportional air pressure in the range from 3 to 15 psi. This unit, shown schematically in Fig. 1-16, is operated by means of a coil suspended in a strong magnetic field. The coil is secured to a pivoted beam which acts as a baffle for the air nozzle. Output current from the amplifier, flowing through the coil, produces a force which moves the beam and changes its position with respect to the nozzle. The nozzle and its associated orifice act as a combined sensing and feedback mechanism.

In operation, an increase in current through the coil forces the beam toward the nozzle, but the resulting restricted airflow develops higher nozzle back pressure. This forces the beam back toward its original position, thus maintaining a balanced condition, with the output air pressure being proportional to the current input to the transducer. A shunting ring provides a means for precise calibration to within a small fraction of a percentage of the span.

The electropneumatic transducer is a plug-in type of unit, designed so that all electrical and air connections are broken when the unit is withdrawn from the case and are remade when the unit is returned to the case.

The transducer can be applied to any situation where it is desired to convert a direct current of 0 to 5 ma into a 3- to 15-psi air pressure. For example, tachometer generators used for speed measurement are available with outputs which can be regulated to the range of 0 to 5 ma, according to rotational speed. For such application, the transducer is supplied in a separate case.

CHAPTER TWO

Pressure-measuring Instruments

WITH THE EXCEPTION OF TEMPERATURE, measurements of pressure and vacuum are the most important measurements made in chemical-processing and manufacturing industries. In fact, it is found in almost any industrial plant that the quantity of pressure gauges far outnumber the other types of instruments, regardless of the kind of product being manufactured. It is therefore essential that the chemical engineer not only be familiar with the available means of measuring and recording pressures but also know how to properly install and maintain pressure-measuring instruments so that maximum service is obtained from them.

The testing and calibrating of pressure-measuring instruments require that the instrument laboratory have available equipment capable of accuracy beyond what is being tested.

PRESSURE MEASUREMENT BY BALANCING AN UNKNOWN PRESSURE AGAINST A KNOWN FORCE

The simplest and oldest method of measuring pressure is by means of an instrument that automatically balances the static pressure being measured against a resisting force whose magnitude can be read directly from the instrument or can be easily computed. This general classification of pressure and vacuum gauges includes instruments that differ widely in appearance, construction, and application, but nevertheless all operate on the same basic principle. Two types usually found in instrument shops or laboratories are used as standards: (1) liquid-column pressure gauge and (2) dead-weight piston gauge.

Liquid-column Pressure Gauge

The Manometer

The liquid-column pressure gauge is usually used to test low-pressure instruments and is referred to in industrial plants as "manometers." These may be constructed of glass or, in rare cases, of metal tubes. They are also constructed in U tubes; that is, the manometer tube is made in the form of a "U." Other manometers may be of the single-tube type with a reservoir which is filled with mercury or some oil, depending upon the range of the gauge to be tested. The pressure applied to the reservoir is also applied to the instrument under test. These manometers usually do not extend in range more than 50 in. of mercury.

Deadweight Piston Gauge

The deadweight piston gauge, or deadweight tester, is usually constructed in various ranges extending up to 1,000 psi. However, deadweight testers are constructed so that they may be used as a standard to test pressure gauges up to 50,000 or even 80,000 psi. The latter, of course, are extreme cases because most processes operate below these ranges.

The deadweight tester of the lower range, illustrated in Fig. 2-1, is no more than a piston supporting weights of known values. Therefore, a pressure gauge may be tested at low ranges, and as weights are added to the piston, the gauge under test is exposed to the same pressure that is generated by turning the crank attached to a piston within a cylinder which is oil-filled and provides a hydraulic pressure to raise the piston.

PRESSURE MEASUREMENT BY DEFORMATION OF AN ELASTIC MEMBRANE

Instruments which use the deformation of an elastic membrane as the primary measuring device are simple, compact, and remarkably free from maintenance problems. For these reasons, the elastic-membrane type of pressure gauge is almost universally used for low- and medium-

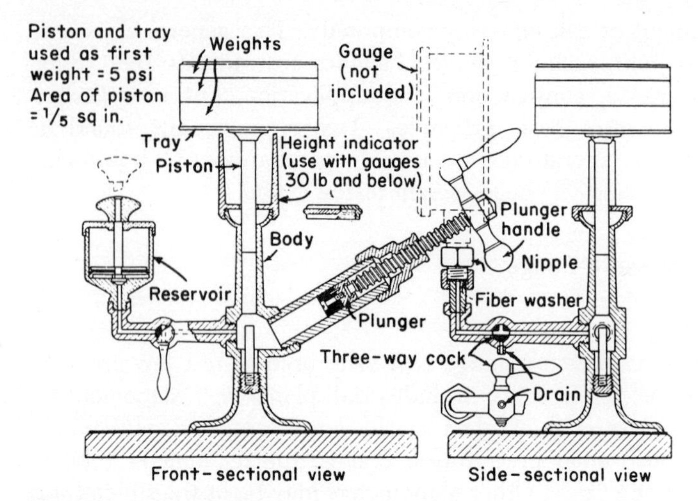

Fig. 2-1 Deadweight pressure gauge tester.

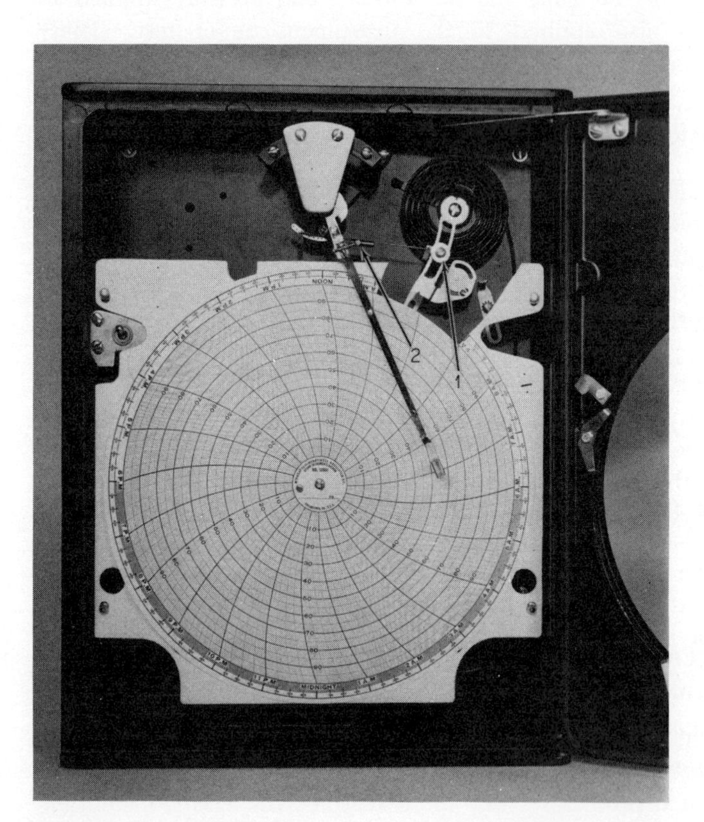

Fig. 2-2 Recording pressure gauge. (*Honeywell, Inc.*)

pressure measurements which are widely used in the field of low-pressure measurement where a large actuating force is not needed.

The three types of instruments in this class are: (1) the Bourdon-tube gauge, (2) the metallic-diaphragm gauge, and (3) the metallic-bellows gauge.

Bourdon-tube Pressure Gauge

The Bourdon-tube type of pressure, or vacuum, gauge is one of the oldest instruments of a strictly industrial type. Pressure gauges of this type were built for the early steam engines long before the chemical-process industries achieved any industrial significance. At present, the Bourdon-tube pressure gauge is numerically the most important instrument used in industry because there are so many applications for pressure measurement and also because the Bourdon-tube pressure gauge is one of the least-expensive instruments.

Figure 2-2 is a recording pressure gauge that uses the Bourdon tube for its primary element. Also shown in Figs. 2-3 and 2-4 are different types of Bourdon tubes.

A pressure transmitter which uses a Bourdon tube and transmits electrically is shown in Fig. 2-5. The output of this electrical pressure transmitter can be transmitted some distance and the chart or scale calibrated so that when the indicator or the recorder is read, it appears as if the pressure is being read at the point of measurement.

The mechanism of a standard Bourdon-tube gauge is shown in Fig. 2-6. The measuring element consists of a metal tube bent into the form of a segment of a circle, having one end fastened to the gauge socket and the other connected through a linkage system to a segment gear and pinion which rotates the pointer shaft.

Bourdon-tube Metals

The metals and alloys used for Bourdon tubes, as well as the heat treatment given the tube, are of vital importance since the performance of the gauge depends to a large extent on these factors. The tube must show little or no tendency to creep or develop hysteresis, even when exposed to maximum pressure for extended periods of time. It must also resist the fatigue stresses imposed upon it because of rapid pulsation of the fluid being measured or excess vibration of the gauge from some other source.

The materials most commonly used for this service are phosphor bronze, alloy steel, stainless steel, beryllium copper, and K Monel. The most widely used metal is bronze, with the bronze tube drawn and bent to the proper arc and then heat-treated to age the metal and prevent the tube from assuming a permanent set under prolonged stress. The bronze tube can be either soldered or silver-soldered to the gauge socket and tube tip, depending upon the pressures and temperatures to be measured.

Bourdon-type Gauge

The combination pressure and vacuum gauge is nothing more than the standard Bourdon-tube gauge with a dial scale graduated in pressure

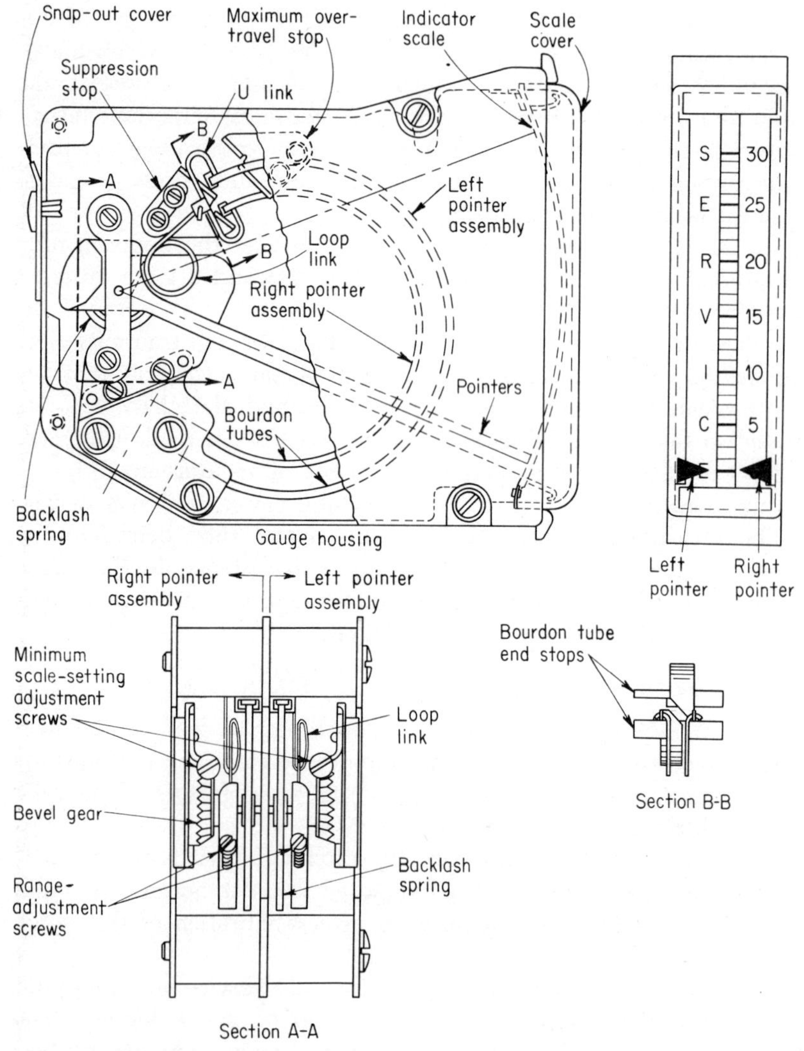

Fig. 2-3 Pressure indicator using a type C Bourdon tube. (*Bailey Meter Company*)

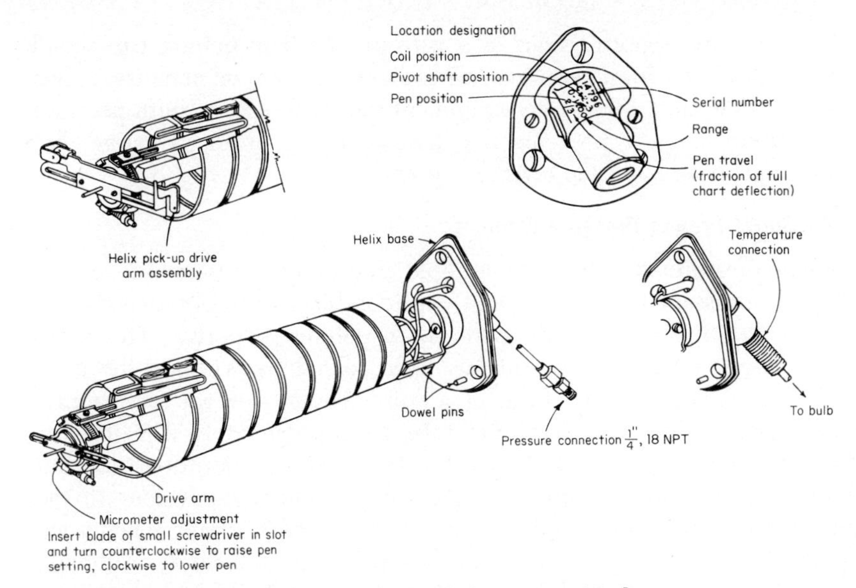

Fig. 2-4 Temperature element of the helix type. (*Bailey Meter Company*)

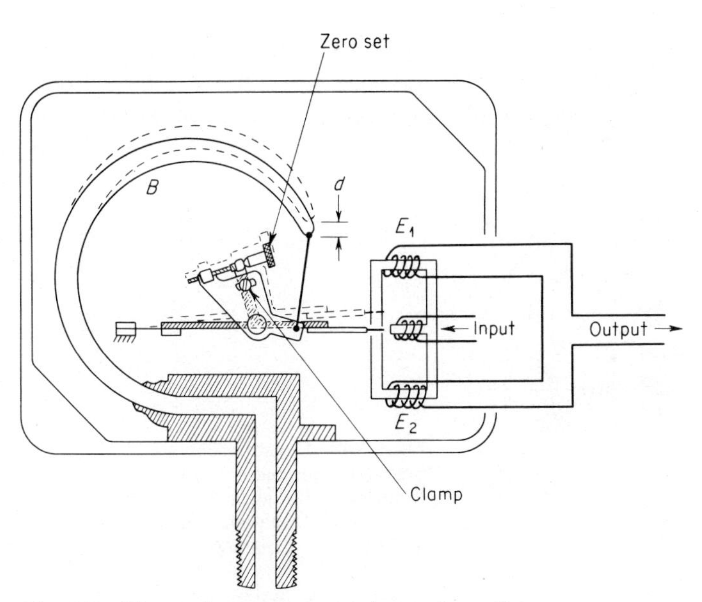

Fig. 2-5 Electrical pressure transmitter using a C type measuring element. (*The Foxboro Company*)

and vacuum on opposite sides of a zero point. The pointer can travel in either direction from zero and will indicate vacuum or pressure. This gauge is convenient for use on equipment that is subject to both pressure and vacuum since it prevents the damage to an ordinary gauge that might result from a sudden reverse in pressure.

The Spiral Type of Pressure Element

When pressure gauges of the simple Bourdon-tube type are recorders, it is necessary to use a tube of considerable size to obtain sufficient travel of the tip to actuate the pen arm or pointer properly. This problem of restricted tip movement has been solved by winding the ordinary Bourdon tube in the form of a spiral having several turns rather than restricting the length of the tube to approximately 270° of arc. This arrangement in no way alters the theory of the Bourdon tube but simply has the effect of producing a tip movement equivalent to the summation of the individual movements that would result when each segment of the spiral is considered as a simple Bourdon tube. Although this construction is more difficult and expensive to build, it has such an obvious advantage for recording or indicating pressure gauges that it is almost universally used for all low- and medium-pressure recorders. This element is also used extensively for recording thermometers. One spiral design is shown in Fig. 2-2.

The Helical Type of Pressure Element

A second variation of the simple Bourdon tube is the helical type of pressure-actuated element. This element is similar to the conventional Bourdon tube, except that the tube is wound in the form of

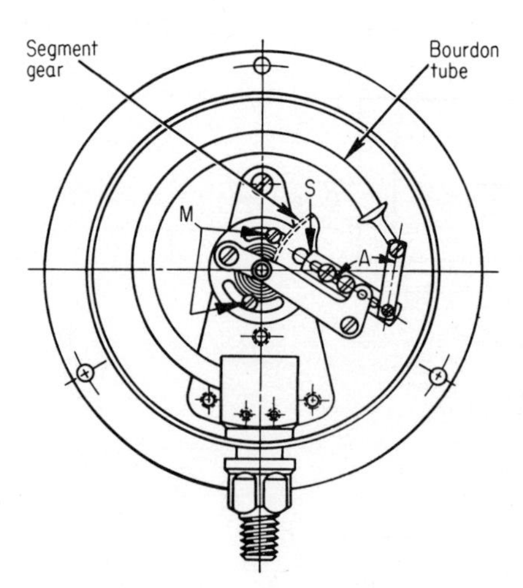

Fig. 2-6 Construction of a pressure indicator using a Bourdon tube as a measuring element. (*Manning, Maxwell & Moore Incorporated*)

a helix having four, five, or more turns. This action increases the travel of the tip considerably and forms a compact unit easily constructed and installed in a recording or indicating pressure gauge. A central shaft is usually installed within the helix and the pen-arm linkage so arranged that it can be driven from this shaft, which in turn is rotated by the tip of the helix. This design transmits only the circular component of the tip movement to the pen arm, which is the component directly proportional to the change in pressure. The helical element is also widely used for recording thermometers. Fig. 2-4 is a measuring element of the helix type.

In addition to recording pressure it is common now to transmit a pressure reading by a pneumatic pressure transmitter, one of which is shown in Fig. 1-12 as a temperature transmitter. This transmitter can be modified and used for pressure transmission also. A pneumatic pressure transmitter can be placed several hundred feet from a recorder; the signal transmitted to the recorder is 3 to 15 psi, which is proportional to the pressure measured by the transmitter.

Metallic-bellows Pressure Gauge

The need for a pressure element that is more sensitive than the Bourdon tube or metal diaphragm for low pressures and that can provide greater power for actuating recording and indicating mechanisms has resulted in the perfection of the metallic bellows. The use of the metallic bellows has been most successful on pressures ranging from 8 oz to 75 lb. The superiority of this type of element over the spiral or helical element can be appreciated by comparison of the power developed by the two. A 2-in. bellows-and-spring element develops 25 times as much power at the same pressure as a seven-turn helix 1¼ in. in diameter.

CHAPTER THREE

Differential-pressure-flowmeter Primary Measuring Elements

MEASUREMENT OF THE FLOW OF FLUIDS is the third most important measurement required of industrial instruments (the other two are temperature and pressure measurements). There are two primary reasons for measurement of flow: (1) to establish ratios of the materials needed in a continuous chemical process and (2) to determine distribution of the material for cost control.

Although the first of these reasons for measurement is of great importance in industry, it is a fact that more flowmeters are used to obtain cost control than for any other purpose. In most chemical industries large quantities of steam are used for process work, and it is virtually impossible to allocate the cost of this steam to the department using it or to discover wasteful use of it, except by means of flowmeters. The use of gas and oil as fuel has also created a significant application for flowmeters; in this service, high accuracy is demanded since the product is usually sold on the basis of the meter recording.

Space does not permit inclusion in this chapter of more than a small part of the data available to the engineer for use in solving flowmeter problems. However, sufficient data are given here to permit the design of flowmeter orifices of the most commonly used type. Other types of meters are studied in sufficient detail to permit easy application of data available from other sources.

INFERENTIAL TYPE OF FLOWMETER

The term "flowmeter" as used herein applies to an instrument for measuring the quantity of fluid flowing in a closed conduit. This will eliminate from the discussion all reference to weirs and the great number of other methods used to measure fluid flow in open channels. These other methods of fluid measurement are important but do not involve instruments of the type covered in this book.

Differential-pressure flowmeters of the inferential type discussed in this chapter are as follows: (1) the venturi tube, (2) the flow nozzle, (3) the orifice plate, and (4) the pitot tube. Each of these types of meters is discussed in theory as well as in mechanical principle.

The inferential type of flowmeter is defined as a meter that obtains a measurement of the flow of a fluid or gas, not by measuring the volume or weight of the medium, but by measuring some other phenomenon that is a function of the quantity of fluid passing through the pipe. The phenomenon usually measured is either pressure differential or velocity in the pipe.

The Venturi Tube

The largest and most significant group of flowmeters is that which uses as an indication of the quantity of flow the difference in pressure resulting from a constriction in the pipe through which the fluid is flowing. These measuring instruments are based on the classic discoveries of the Italian Venturi who, after valuable research in 1797, announced that fluids under pressure gain speed and lose head in passing through converging pipes and that the reverse is true for fluids passing through diverging tubes. This principle was not used in a fluid-measuring device for more than one hundred years after it was known as a scientific fact. Clemens Herschel, the first to use this principle as a means of measuring flow, proved by a series of elaborate tests that a remarkably constant relation in a convergent-divergent tube existed between the difference in pressure at the inlet and the contracted sections and the rate of flow. Herschel named his instrument a "venturi-type" primary element.

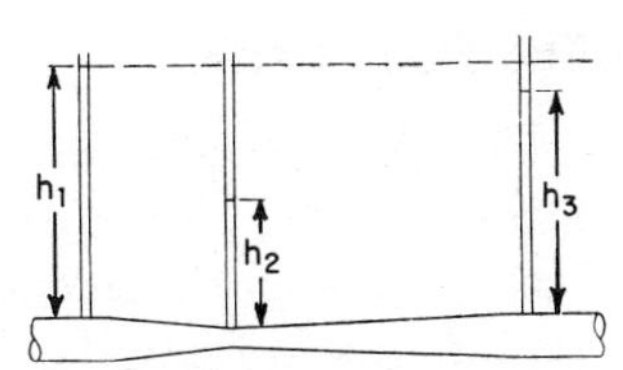

Fig. 3-1 Schematic diagram of the venturi tube.

A schematic diagram of a venturi tube is shown in Fig. 3-1. With a fluid flowing through this type of constriction, the static pressure (represented by the water columns) drops sharply as the throat of the tube is approached and recovers most of the original static head on the downstream side of the tube. This phenomenon discovered by Venturi is the basis of most differential-pressure meters.

Pressure Recovery in the Venturi-type Primary Element

The pressure distribution in a venturi-type primary element can be observed in Fig. 3-1 which shows that as the venturi returns to the normal pipe size, the static pressure returns to a value nearly as great as the upstream pressure. If the venturi is carefully constructed with smooth sidewalls and with a correct slope to and from the throat, it is possible to recover as much as 85 percent of the original static pressure. High-pressure recovery is one of the advantages of this type of meter.

General Design of Venturi Tubes

The venturi tube is built of cast metal and machined to close tolerances by the manufacturer. The size may vary from a few inches to 60 in. in diameter and 40 ft in length with a flow capacity of 80,000 gal per min. Larger sizes are built in several sections and bolted together. A typical venturi tube is shown in Fig. 3-2. The upstream

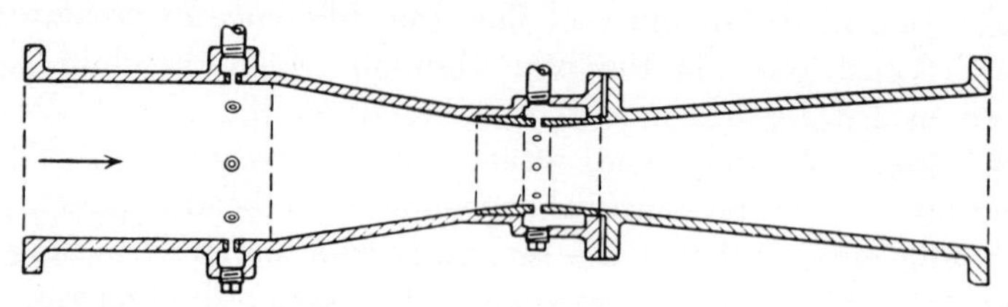

Fig. 3-2 Typical venturi tube construction.

cone of the venturi usually forms an angle of approximately 35° with the axis, and the downstream cone forms an angle of approximately 7°.

The Flow Nozzle

The flow nozzle, an adaptation of the venturi tube, is often used to measure flow by means of differential-pressure measurements. This type of primary element, shown in Fig. 3-3, consists essentially of a venturi tube without the diverging section.

Developments in flow nozzles in recent years have improved some measurements and have produced a low-differential pressure loss as the flow passes through the nozzle. (The Dall tube shown schematically in Fig. 3-4 is one.) The curves in Fig. 3-5 are self-explanatory. At

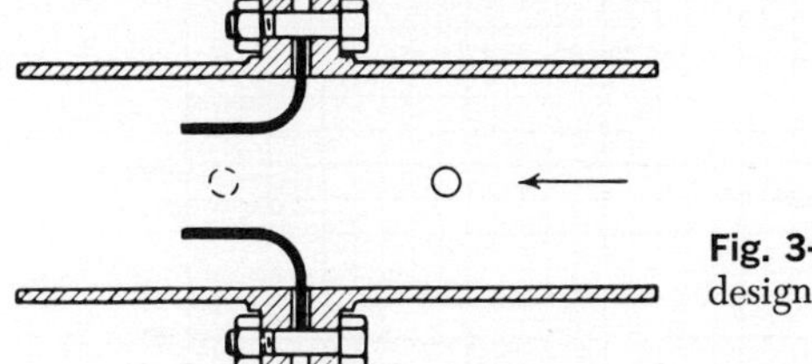

Fig. 3-3 A typical flow nozzle design.

times the Dall flow nozzle is preferred over other primary elements if horsepower is at a high cost or if a small additional loss in pressure would overload a compressor. Therefore, when a control system is being designed where large lines are involved, it is well to consider that the Dall flow tube, which operates satisfactorily if properly cared for, may be preferred.

The Orifice Plate

The orifice plate is by far the most widely used type of primary element for differential-pressure meters. The theory of this primary element is the same as that of the venturi tube and the flow nozzle.

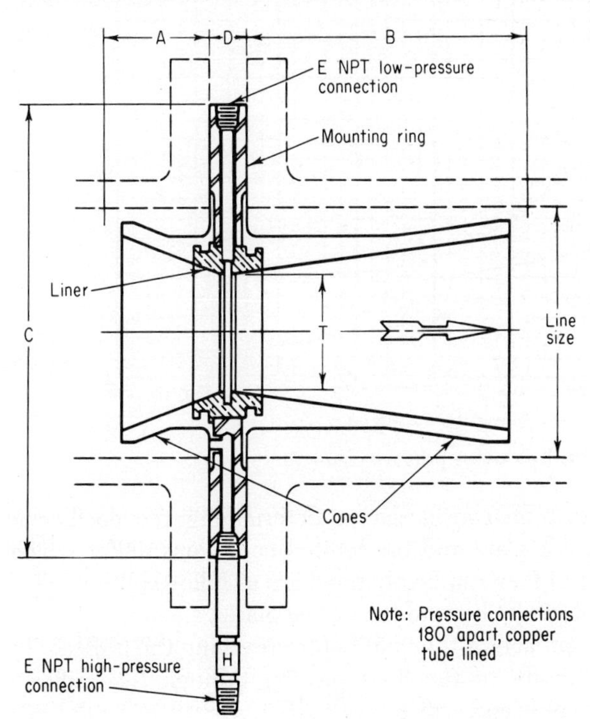

Fig. 3-4 Schematic of a Dall tube.

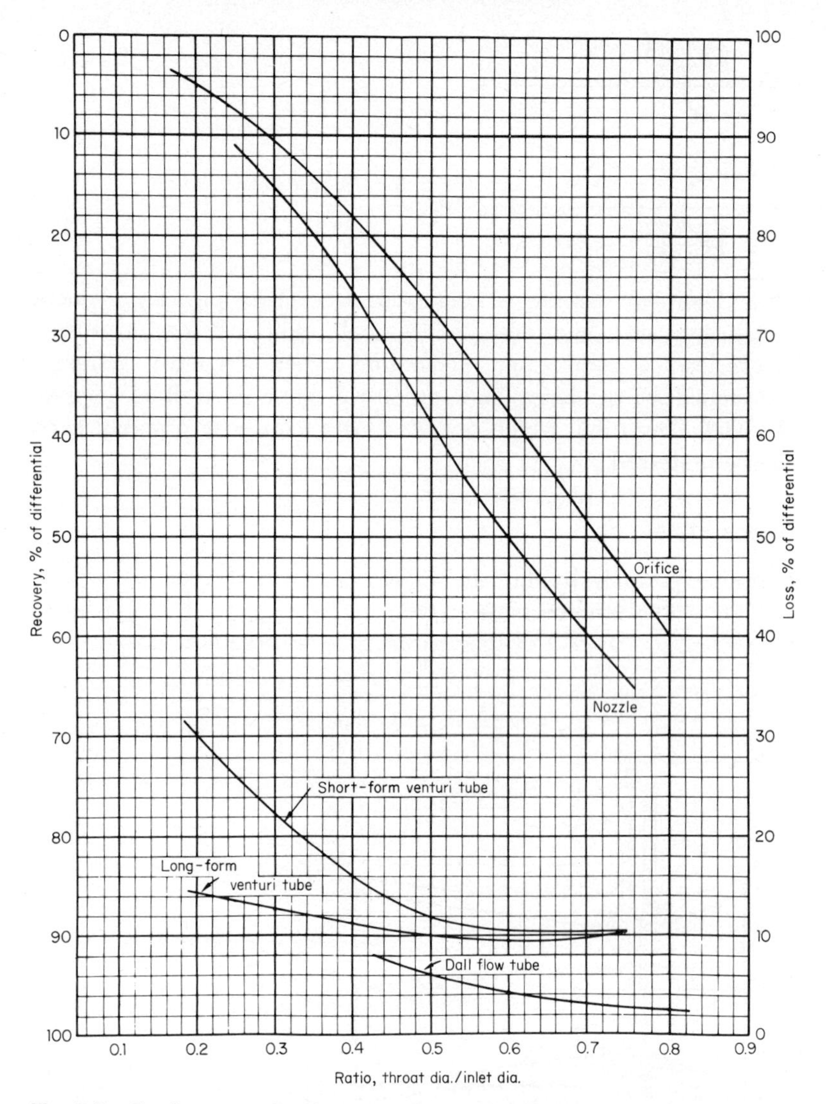

Fig. 3-5 Total pressure loss by various primary elements.

The factors that must be considered in determining the coefficient to be used with an orifice plate and the limited conditions under which accurate measurements of flow can be obtained are as follows:

1. The contraction of the jet beyond the orifice plate
2. The velocity of approach of the fluid before reaching the orifice
3. The change in density of the fluid passing through the orifice
4. The friction in the pipe and orifice
5. The problem of viscous flow
6. Deviations from the perfect gas law

7. Correction of coefficient for the Reynolds number
8. Pulsating flow

Each of these factors will now be considered in detail.

Contraction of the Jet beyond the Orifice

In the case of the venturi tube and the flow nozzle, the flow follows the contour of the venturi tube (if properly designed), and the throat area is considered as the area to be used for flow computations. For the orifice plate, the area of the jet to be used in the flow equations is not the orifice area but a smaller area located beyond the plate. This is a familiar hydraulic phenomenon illustrated clearly in Fig. 3-6. In solving the equation of flow, we must take this factor into account in the coefficient of flow.

The contraction of the jet also influences the location of the vena contracta, or point of greatest pressure drop, and causes the low-pressure tap to be moved downstream from the orifice plate.

Velocity of Approach to the Orifice

In the development of the equation of flow from the hydraulic formula the velocity of approach of the fluid to the orifice is assumed to be negligible, a factor taken into account in the flow coefficient. However, this variable accounts for a considerable part of the variation in the flow coefficient with variations in the ratio of orifice diameter to pipe diameter.

Change in Density of Fluid Passing through Orifice

The development of the flow equation from the simple hydraulic formula assumes that the density of the fluid does not change as it

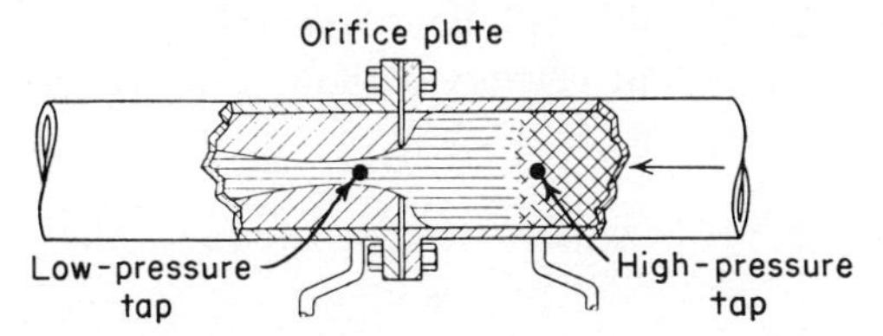

Fig. 3-6 Contraction of the jet beyond the orifice.

passes through the orifice. For a gas this is never exactly true, although it may be nearly so where the gas being measured is under high pressure and has a low-pressure differential. Under ordinary conditions the gas becomes somewhat less dense and drops in temperature as it passes through the orifice.

Fortunately, the change in density of a gas is small enough under the usual conditions of gas measurements not to destroy the commercial accuracy of the readings. The effect of this change in density has been found by test to produce a variation of less than 1 percent from the average of the upstream and downstream densities if the following rule is applied: The differential pressure in inches of water should not exceed the absolute static pressure in pounds per square inch.

If it is necessary to use a differential pressure greater than that allowed by the foregoing rule, it is possible to use a static pressure that is an average of the upstream and downstream pressures. This makes it possible to measure flows of higher velocity without serious error. However, the foregoing rule should always be followed if at all possible. The magnitude of the error resulting from a change in gas density can be seen in Fig. 3-7 which shows a chart of the Foxboro Company, giving the correction obtained by the use of the average static gas pressure.

Frictionless Flow

The theoretical equation of flow through an orifice assumes that there is no friction loss in the system. Obviously this is not true in the most refined design of flow nozzle and is certainly not true in an orifice plate. This factor is always taken into account in the orifice coefficient and is a part of the problem that does not lend itself to theoretical analysis. A considerable part of the reduction in flow from the theoretical value is due to this cause.

Viscous Flow

It is well known that flow of fluids in a pipeline exhibits three distinct flow characteristics, depending upon the velocity of the flow, the specific gravity of the fluid, the absolute viscosity of the fluid, and the pipe diameter. Tests have shown that a fluid will flow smoothly through a pipe without turbulence up to a certain velocity of flow. Above this

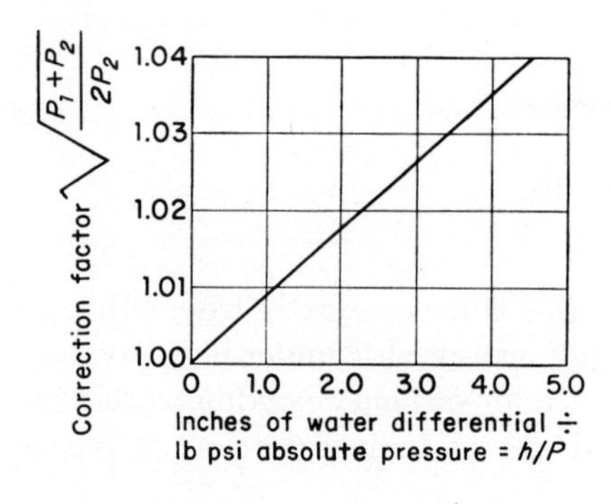

Fig. 3-7 Correction obtained by use of average static pressure $P_1 + P_2/2$. (*The Foxboro Company*)

velocity the flow becomes turbulent, and eddies are set up that cause a greater pressure drop in the line and a change in the distribution of velocity throughout the pipe.

Figure 3-8 shows the distribution of velocity in a pipeline under the

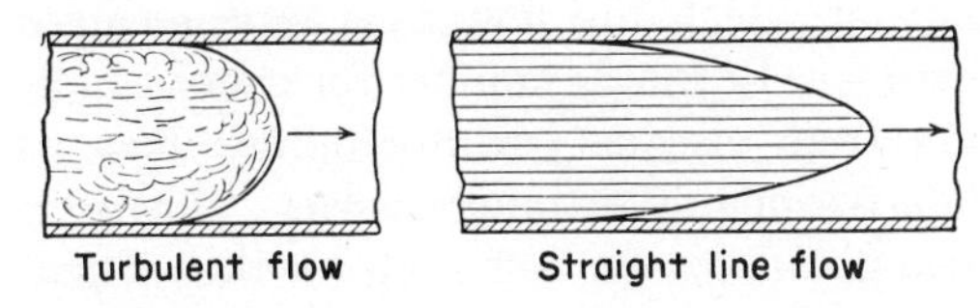

Fig. 3-8 Distribution of velocity in pipe lines.

condition of turbulent flow and under the condition of viscous flow. It can be seen that for viscous flow, the velocity at the center of the pipe is approximately twice the mean velocity; for turbulent flow, the velocity at the center is much less. This change in flow characteristics has a certain influence on the orifice coefficient, particularly if the ratio of orifice diameter to pipe diameter is large and if the pipe size is small. Fortunately, streamline (or viscous) flow is a source of serious inaccuracy only under rather unusual conditions. It is evident that streamline flow causes an error on the high side since the velocity at the center of the stream is increased.

Deviations from the Perfect Gas Law

It is assumed that the gases flowing through an orifice-type primary element obey Boyle's law which states that at a constant temperature the density of the gas varies with the pressure. Most gases do not follow this law exactly but tend to increase in density more rapidly than would be the case if the gas were ideal. This phenomenon is more pronounced at high pressures; and since the error varies in magnitude with the pressure, temperature, and kind of gas being measured, it is difficult to evaluate the error except for specific flow conditions. The error is not serious because if the density is greater than that obtained from the gas law, it will represent a larger number of cubic feet of gas at standard conditions but will also have produced a greater pressure differential for the given velocity. Hence to a certain extent, the error cancels out.

Reynolds-number Correction*

It is possible to obtain flow measurements with an orifice plate within

* L. K. Spink, *Principles and Practice of Flow Meters Engineering*, 9th ed., the Foxboro Company, Foxboro, Mass., p. 191.

the limits of commercial accuracy without the use of any correction for Reynolds number. However, if the highest possible accuracy is required, it is necessary to make a correction for this characteristic of the flow.

The Reynolds-number criterion can be described as a numerical value indicative of the critical velocity at which the flow in a pipe changes from viscous to turbulent. There is a Reynolds number for every condition of flow, and the transition from viscous (or streamline) flow to turbulent flow occurs at a certain reasonably definitive valve. However, a Reynolds number may be computed for flows far above the critical velocity.

The flow through an orifice varies with the value of the Reynolds number obtained for any particular set of flow conditions, and research carried on by the American Gas Association, the American Society of Mechanical Engineers Committee on Fluid Meters, and the U.S. Bureau of Standards has correlated the Reynolds number with the flow coefficient. As a result it is possible to use this value as a means of increasing the accuracy of flow measurements. The correction will be made in all flow equations described in the following section on orifice-type primary element computations.

The size of the orifice plate may be anything desired. However, the greatest amount of experimental data is available in the range of pipe sizes between 1½ and 12 in., although much larger-size pipes are sometimes fitted with orifice plates. Orifice plates are also used with smaller pipe sizes, but in this case data must be supplied by the manufacturers for computing the flow, or if made in the plant shop, the orifice must be calibrated before it is used. For the smaller sizes this is a comparatively easy task since only small weigh tanks or gas tanks are needed to determine the coefficient. For pipes larger than 12 in. the problem of obtaining a calibration in the field is next to impossible because of the size of the equipment needed to determine the coefficient. Where possible, it is best to design the orifice for a pipe within the range of 1½ to 12 in. in diameter. However, pipe sizes up to 30 in. in diameter are practical.

Location of the Pressure Taps

Three arrangements of pressure taps, each having advantages in certain cases, are used with the orifice plate. They are: (1) vena-contracta taps, (2) flange taps, and (3) pipe taps.

The first tap arrangement is shown in Fig. 3-9 and requires that the tap on the upstream side be located approximately one pipe diameter above the orifice face and the downstream tap a varying distance, depending upon the orifice size. Location of the downstream tap is such

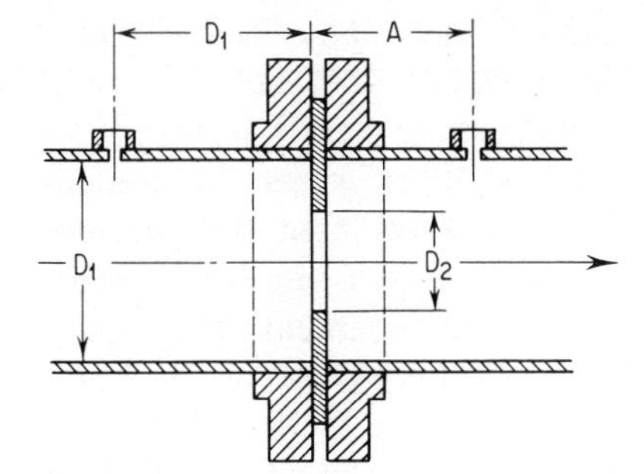

Fig. 3-9 Vena contracta taps.

that it measures the pressure at the vena contracta (or point of minimum static pressure). This tap arrangement produces the maximum pressure differential for a given flow and when properly installed is probably slightly more accurate than the other tap arrangements since there is a greater pressure differential for each increment of change in flow. The disadvantage of this kind of tap arrangement is that the taps must be tapped or welded into the pipe accurately and that, under certain conditions, the downstream tap may not clear the pipe flange, in which case a correction must be made to compensate for the location of the tap farther downstream.

The location of flange taps is shown in Fig. 3-10. This kind of tap

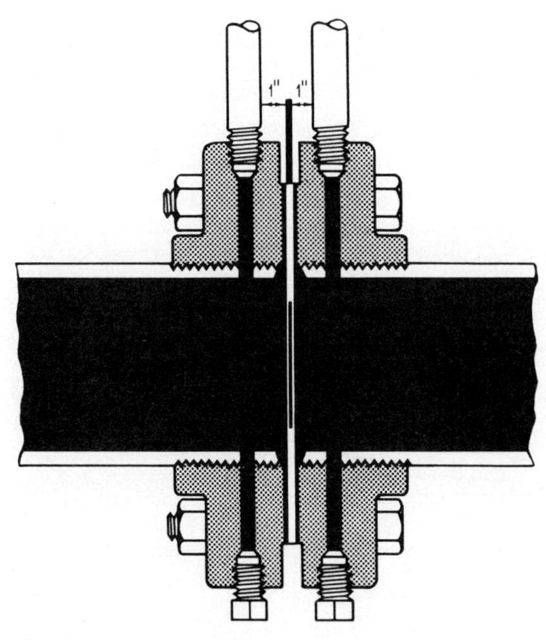

Fig. 3-10 Flange taps.

arrangement locates both taps in the pipe flanges at a fixed distance from the orifice plate, usually 1 in. The distance does not vary with the size of pipe or with the ratio of orifice diameter to pipe diameter.

The location of the taps in the flange sacrifices a certain amount of pressure differential, but study of Fig. 3-11 shows that the pressure drop at the downstream flange is great enough to give significant readings. Hence for most installations, this tap arrangement is as good as the vena-contracta-tap arrangement. Its advantages are obvious and are listed as follows:

- The unit does not require drilling of the pipe, with the attendant possibility of error in location of the taps.
- The unit is compact and may be installed without disturbing the piping.
- The orifice may be changed to a different size without the necessity of applying a correction for mislocation of taps, as is the case with vena-contracta taps.
- The unit may be purchased with the taps accurately located by the manufacturer.

These advantages make this type of installation desirable in most cases where only a few meters are used and where an experienced construction department is not available for the installation of vena-con-

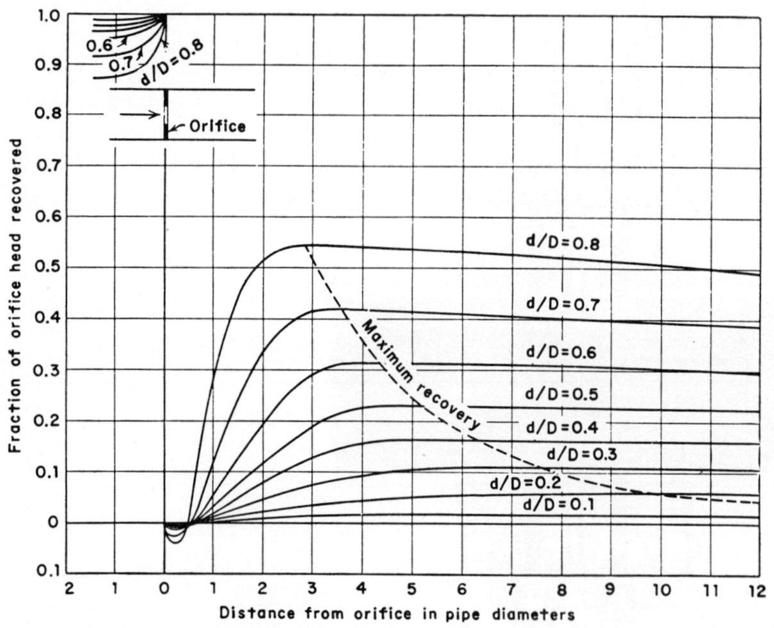

Fig. 3-11 Pressure distribution in the vicinity of an orifice plate. (*The Foxboro Company*)

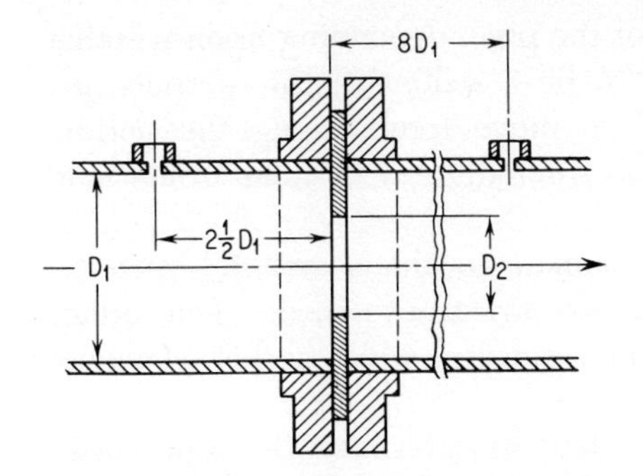

Fig. 3-12 Pipe taps ($2\frac{1}{2}D_1 \times 8D_1$).

tracta taps. If large numbers of meters are used and competent help is available, the vena-contracta-tap arrangement can usually be installed more cheaply than the flange type because special flanges are not needed.

The arrangement of pipe taps is shown in Fig. 3-12; they are usually located two and one-half pipe diameters upstream from the orifice plate and eight pipe diameters downstream from the orifice plate. These taps do not vary with the size of the orifice plate. It can be seen that they measure only the permanent pressure drop across the orifice since they are located approximately at the point of maximum pressure recovery as shown in Fig. 3-12.

The Segmental Orifice Plate

A variation of the simple circular orifice plate is shown in Fig. 3-13. It consists of a straight-edged baffle installed at a pipe flange with the usual arrangement of pipe taps. This type of orifice produces a pressure differential nearly as large as that of a circular orifice of the same area and may be used to advantage under certain conditions. The segment

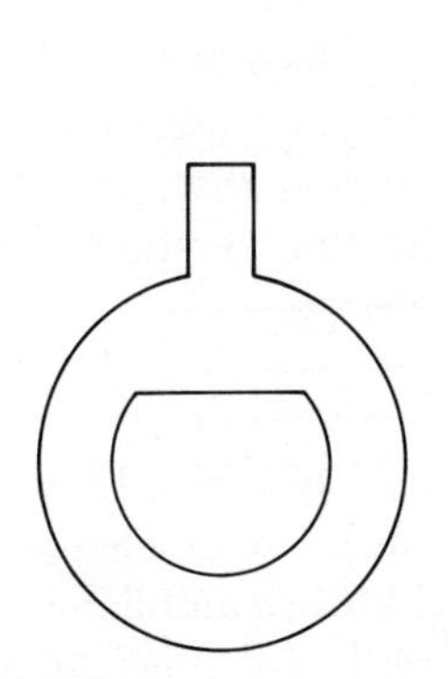

Fig. 3-13 Segmental orifice plate. (*Bailey Meter Company*)

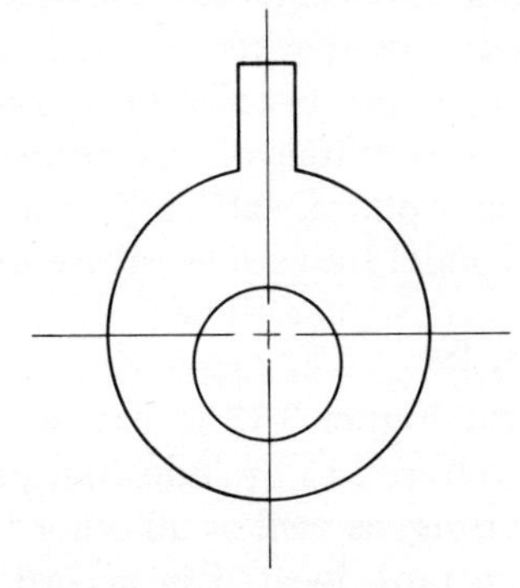

Fig. 3-14 Eccentric orifice plate. (*Bailey Meter Company*)

is installed at the top or the bottom of the pipe, depending upon whether gas or air is to be unobstructed or to be a sediment; this permits any sediment or other foreign material to move freely along the bottom of the pipe, whereas it would be obstructed by a circular orifice and would build up on the upstream side.

Figure 3-14 illustrates an orifice similar to the segmental plate in design, except that it is round and eccentrically located. This orifice serves the same purpose as the straight segmental orifice and is somewhat easier to construct.

Both these orifices require special data to determine the flow coefficient, and these are usually supplied by the manufacturer. However, it should be remembered that these orifices are most advantageous where fluids carrying solids in suspension are to be measured; in such cases the specific gravity of the fluid may be unknown or variable. In this case a calibration of the orifice in service is desirable.

The Quick-change Orifice Plate

It will be shown in a subsequent section on the design of flowmeter mechanisms that the range of flow through which accurate readings can be recorded is limited for any particular size orifice. Hence, if the rate of flow varies widely, it is necessary to provide a means of adopting the metering equipment in one of the following ways:

1. Provide a bypass meter to take the flow at low velocity.
2. Provide an easy way of changing orifices when the flow varies.
3. Provide a method of changing the flowmeter mechanism to record accurately either a high- or a low-pressure differential.

The first method is simple and practical where the installation warrants the expense of two separate flowmeter units. In operation this system is quick and positive—all that is required if the flow drops to a certain low point is to open one valve and close the other.

The second method is used where a single meter is desired. In this instance the accuracy of the installation is maintained by inserting an orifice to suit the flow conditions. An orifice installation that permits the changing of orifice plates with pressure on the line is shown in Fig. 3-15, which is a Daniel Industries, Incorporated, design.

Daniel Industries, Inc.

PRINCIPLE OF DESIGN: Figure 3-15 is a cross-sectional view of a Daniel senior orifice fitting, referred to by industrial people as a "quick-change" orifice fitting. The fitting, as well as all other Daniel fittings and flanges, has two sets of flange taps, located in accordance with the latest AGA recommendations. The flange-tap meter connections are ½-in. NPT. The internal tap holes also meet the latest AGA recommendations.

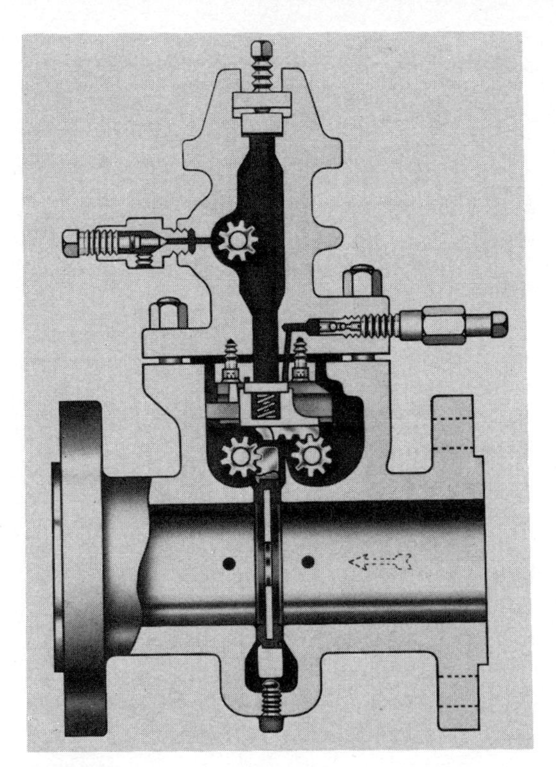

Fig. 3-15 Quick-change orifice. (*Daniel Industries, Inc.*)

Line-bore tolerance for 2- and 3-in. sizes is ±0.003-in.; for 4-, 6-, 8-, and 10-in. sizes, ±0.004-in.; and for 12-in. and larger sizes, ±0.005-in. Unless otherwise specified, all fittings are furnished with standard internal line bore as listed in dimensional tables.

On sizes 2 to 14 in. inclusive, single operating shafts are standard on the left side of the fitting when looking downstream. On sizes 16 in. and larger, operating shafts on both sides of the fitting are standard.

All senior orifice fittings are composed of two independent compartments separated by a hardened stainless-steel slide valve. Shown is the slide valve in closed position and the orifice plate concentric in the line of flow. The slide valve cannot be closed unless the orifice is concentric to the bore of the fitting.

The Daniel patented top closure cannot be installed unless the plate carrier is in place. Only a few turns of the speed wrench supplied with the fittings are required to remove or replace the clamping and sealing bars. Setscrews always remain in the clamping bar, adding greatly to speed and ease of operation.

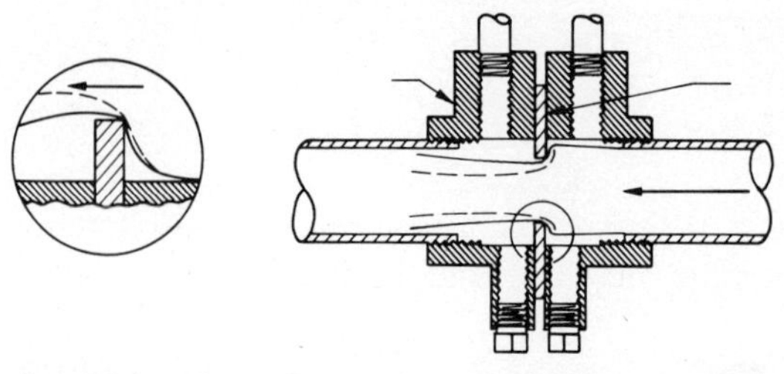

Fig. 3-16 Characteristic of rounded edge of orifice. (*The Foxboro Company*)

The plate carrier is raised and lowered by the double rack-and-pinion mechanism with power applied through the speed wrench. This method provides the quickest means of operation with the least amount of effort and assures positive control of the plate carrier at all times.

Maintenance of the fitting is simple because all essential parts can be replaced or removed without having the fitting removed from the pipeline. Many gas and oil companies have made this fitting standard because of ease of operation; excellent sealing characteristics; standard stainless trim for sour gas; expert machining of body seats; meter-tube fabrication; wide range of sizes, pressures, and metal trims; and body design which can be raised-face flanges on each end, flanged on one end and welded on the other, or flanged on both ends, thereby making it adaptable for many installations.

COMPARISON OF DIFFERENTIAL-PRESSURE-METER PRIMARY ELEMENTS

The three most common primary elements used with flowmeters have been described in the preceding paragraphs without specific reference to the particular advantages and disadvantages of each type. A better idea of the proper type of primary device to use for a particular installation can be obtained from the following analysis of the advantages and disadvantages of each type of orifice.

The Venturi Tube

Advantages

High accuracy: The venturi tube, when carefully made and calibrated, is probably the most accurate unit available.

High efficiency: The pressure recovery of the venturi tube is greater than in any other primary element, except the Dall tube, and should be used where this is important.

Accuracy with high ratios of orifice diameter to pipe diameter: The venturi tube gives accurate reading with a ratio of d/D greater than 0.75 and should be used when high ratios are necessary.

Resistance to abrasion: the venturi tube has smooth curves and resists wear effectively.

Does not catch dirt or sediment: Because the tube is smooth, there is no obstruction to catch foreign material and in this way introduce a source of error.

Disadvantages

Costly: The venturi tube is expensive to make and must always be purchased from a manufacturer with a proved coefficient.

Cannot be changed easily: Owing to its high cost it is not desirable for use where variations in flow require changes in orifice size.

Large size: The venturi tube is long and consequently awkward to install.

The Flow Nozzle

Advantages

High efficiency: The pressure recovery is nearly as great as that of the venturi tube. The Dall tube has an advantage in this case. (Refer to curves shown in Fig. 3-5 for a comparison of the pressure recovery of the various primary elements.)

Accuracy with high ratios of orifice to pipe size: The flow nozzle operates on d/D ratios greater than 0.75.

Good resistance to wear.

Disadvantages

High cost: The flow nozzle is not so expensive as the venturi tube but is still high compared with the orifice plate.

The Orifice Plate

Advantages

Low cost: The orifice plate can be purchased cheaply or made in the plant shop.

Little room required: The orifice plate may be inserted in available flanges with a minimum of changes.

Easily interchanged: Orifice size may be quickly changed to meet a new flow condition.

Can be constructed with variable orifice.

Complete data available for computing flow: Flow easily computed for any size of plate.

Disadvantages

Low efficiency: Has a greater pressure loss than other type.

Poor accuracy on high ratios of orifice size to pipe size.

Subject to wear: Abrasive material will wear the sharp edge of the plate and cause inaccuracies. Refer to Fig. 3-16 for conditions affecting accuracy of measurement.

Obstructs flow of dirt or sediment in the pipe.

Somewhat fragile: May be damaged by water-hammer or foreign material flowing in the pipe.

Pitot Tube

The fourth type of differential-pressure-flowmeter primary element is the pitot tube. The third and most important kind is the orifice type which has just been covered in detail. The pitot tube is a valuable measuring tool when used properly on the right application. It has two major applications: (1) measurement of flow where a temporary installation is satisfactory and (2) measurement of flow where the size of the pipe or duct is too great to make an orifice installation practicable.

A major objection to the pitot tube which confines its practical application in most cases to a test instrument is that it cannot be used for any length of time in a fluid containing dirt or solids. Any foreign material either clogs the tube completely or builds up on the exposed surface to such an extent that it seriously changes the calibration of the instrument. As a result of this difficulty a pitot tube must be thoroughly cleaned before measurements are taken, and the instrument is not usually reliable as the primary element for a recording flowmeter.

Theory of the Pitot Tube

There has been a large number of variations in the design of the pitot tube, and in most cases the manufacturer has developed a formula to meet the requirements of the particular pitot tube constructed. However, in all cases the fundamental basis for the flow measurement is the value of the difference between the static and the total of static and velocity head at the same point in the stream of flow. The typical arrangement of the pitot tube used to obtain this value of velocity head is shown in Fig. 3-17.

The theoretical equation for the velocity head measured by a pitot tube is

$$V = \sqrt{2gh}$$

where V = velocity of the fluid, ft per sec
h = velocity head as measured with a manometer
g = acceleration constant of gravity

This is the familiar law of a free-falling body. The validity of the

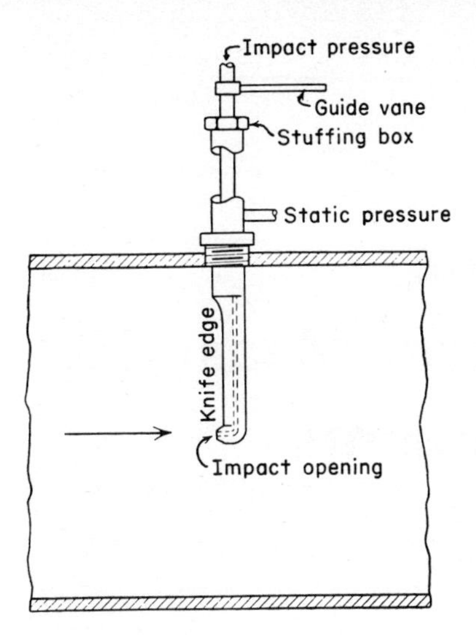

Fig. 3-17 Typical pitot tube.

equation as the basic formula for pitot tubes has been well established experimentally.

For the orifice type of primary element, the theoretical equation for the velocity of flow as given above must be corrected by an empirical coefficient C in order to represent the actual flow obtained in practice. This coefficient must take into account the friction of the fluid, deviations in the stream of flow caused by the pitot tube itself, and errors in the measurement of the static pressure obtained by the usual arrangement of static-pressure connection.

Likewise the factors of viscosity, specific gravity, Reynolds number, moisture or quality, etc., all affect the flow as measured by a pitot tube in much the same way as they affect the measurement of flow by means of the orifice meter. Manufacturers of pitot tubes usually use exactly the same equations for measuring flow with pitot tubes and orifice plates, changing the flow coefficient C only to correct the equation for the difference to flow represented by the same pressure differential in an orifice plate and a pitot tube.

CHAPTER FOUR

Differential-pressure-flowmeter Secondary Measuring and Recording Elements

THE PRIMARY MEASURING ELEMENTS studied in the preceding chapter were designed to produce a certain pressure differential for a certain rate of flow through a pipe, and the methods used to evaluate the flow for any particular pressure differential were explained. It is now desirable to cover the various mechanisms that are used to measure and record the pressure differential obtained from the primary flowmeter elements.

RECORDING FLOWMETER

In the majority of cases where a flowmeter is needed it is important not only to be able to read the rate of flow at any time but also to obtain a record of the rate of flow over a period of time. Without this chart record it is difficult to find the peak load on a system since, if readings are taken as often as every 15 min, the maximum or minimum flow may deviate widely from that noted by an operator. This is particularly true where the flow fluctuates suddenly and over a considerable range. Even if the flow does not vary suddenly, it is not always convenient to have an operator take the necessary time to make frequent entries on a log sheet. Furthermore, in many cases the flowmeter is not equipped with a mechanical integrator, and the only way of determining the total flow over a period of time is to compute the average.

Bailey Meter Company

principle of design: Figure 4-1 shows a cutaway view of an inverted bell-type meter of the Bailey Meter Company design. The Bailey meter employs a Ledoux bell which is designed to extract the square root of the differential pressure across such primary elements as orifice plates, flow nozzles, or venturi tubes and to record or indicate it on a linear chart or scale.

Since the quantity of a fluid flowing through a primary element such

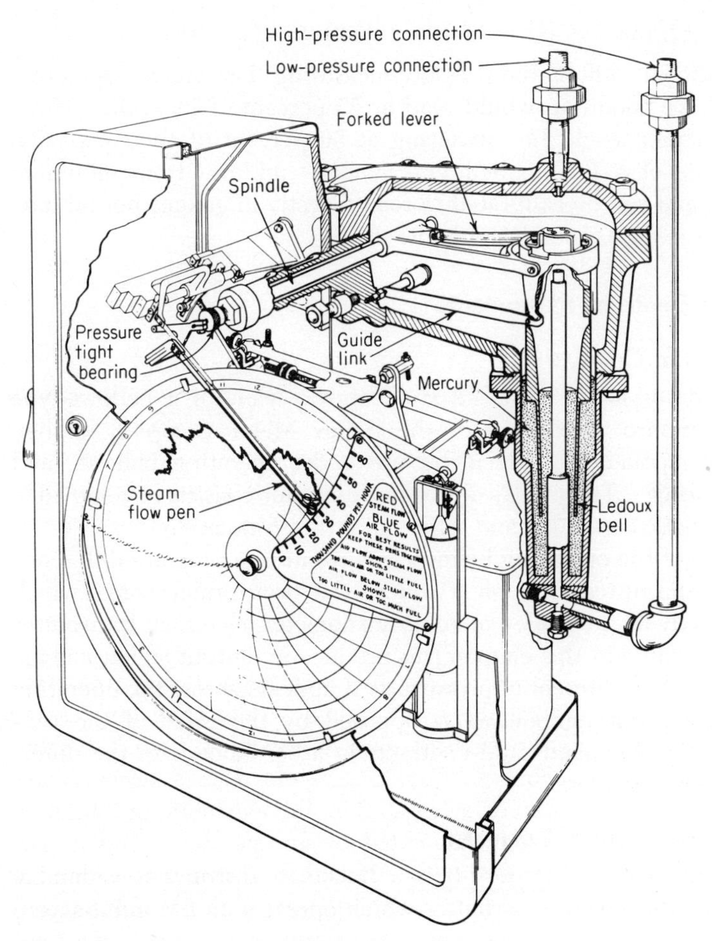

Fig. 4-1 Dual-type Ledoux-bell meter. (*Bailey Meter Company*)

as an orifice, flow nozzle, or venturi varies as the square root of the differential pressure across the primary element, a meter designed to measure the differential pressure would have a compressed scale in the low range and an expanded scale in the high range. This is undesirable where widely fluctuating flows are to be measured. This undesirable feature is overcome by the Ledoux-bell meter because the differential pressure in inches of water required to move the pen or pointer a given percentage of the chart or scale is equal to that percentage divided by 100 percent, squared, times the maximum range of the meter in inches of water. For instance, if a 50-in. Ledoux-bell meter has a 50 percent scale reading, what differential pressure would exist across the Ledoux bell?

$$(50/100)^2 \times 50 = 12.5\text{-in. differential pressure}$$

A conventional differential-pressure-measuring instrument operating under the above condition would read at 25 percent of its scale. However, both meters would be measuring at 50 percent of their capacity. Thus the scale of a Ledoux-bell-type meter is uniform from minimum to maximum and can be calibrated to read directly in gallons per minute, cubic feet per minute, pounds per hour, etc.

Transmitting Electrical Instruments

Bailey Meter Company*

PRINCIPLE OF DESIGN: Figure 4-2 is a cutaway view of a bellows-type differential-pressure transmitter of the Bailey Meter Company design. The major components of the meter are body, operating bellows, and calibration spring. The electrical transmitting unit consists of a movable core, nonmagnetic core tube, and differential transformer.

Any motion of the operating bellows as a result of a pressure difference across it is transmitted through a rod to the transformer core (which functions exactly as the Bailey mercury-manometer, electrical transmitter discussed elsewhere in this chapter) to produce an output signal voltage that is linear to the differential pressure as it appears across the operating bellows. The operating bellows is opposed by the force of the calibrating spring and expands and contracts in accordance with the difference in pressure across it.

Beckman Instrument Company

PRINCIPLE OF DESIGN: Figure 4-3 is a Beckman Instrument Company VDP differential-pressure transmitter which operates on the null-balance

* Grady C. Carroll, *Industrial Process Measuring Instruments*, 1st ed., McGraw Hill Book Company, New York, 1962, pp. 39–40.

vector principle to translate the force of a differential-pressure-sensing element into a 3- to 15-psi air pressure. The differential pressure-sensing element consists of two body forgings between which a measuring diaphragm is clamped, a combination bellows seal and pivot unit, and an appropriate linkage system for transmitting the motion of the measuring diaphragm to the pneumatic transmitting unit.

Major components of the unit are shown schematically in Fig. 4-4. When the unit is in operation, with no differential pressure applied to diaphragm R, input force is zero and diaphragm lever K and tension member M are in a static state. The vector zero-adjustment spring, however, produces a fixed force in the direction of Q. This force moves arm P which is attached to the nozzle baffle; the nozzle produces a pressure drop on the reversing air relay in response to the baffle. The relay, in turn, produces an increasing output pressure to the feedback diaphragm and receiving instrument. The increasing output pressure acts on the feedback diaphragms to produce an increasing net force in the direction of C until force Q is balanced. If we assume that

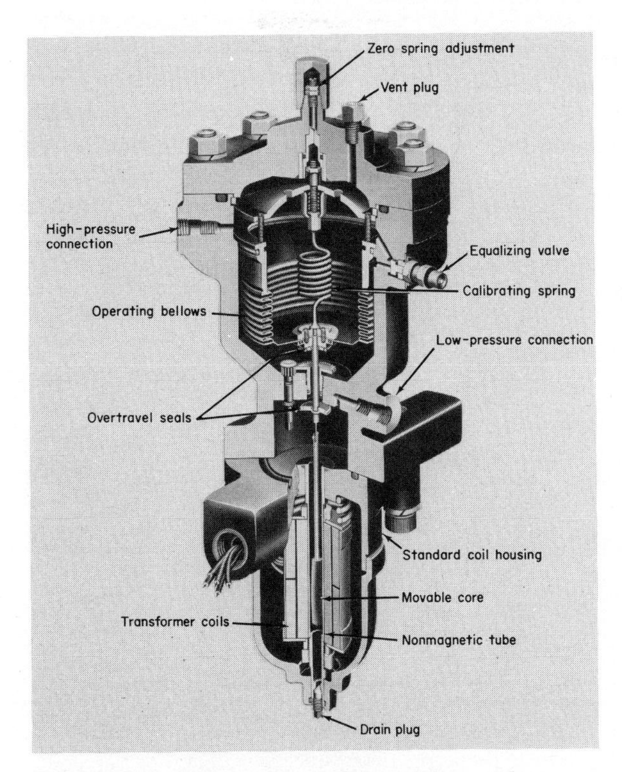

Fig. 4-2 Dry bellows electrical-type differential-pressure transmitter. (*Bailey Meter Company*)

Fig. 4-3 Differential-pressure transmitter, pneumatic type. (*Beckman Instrument Company*)

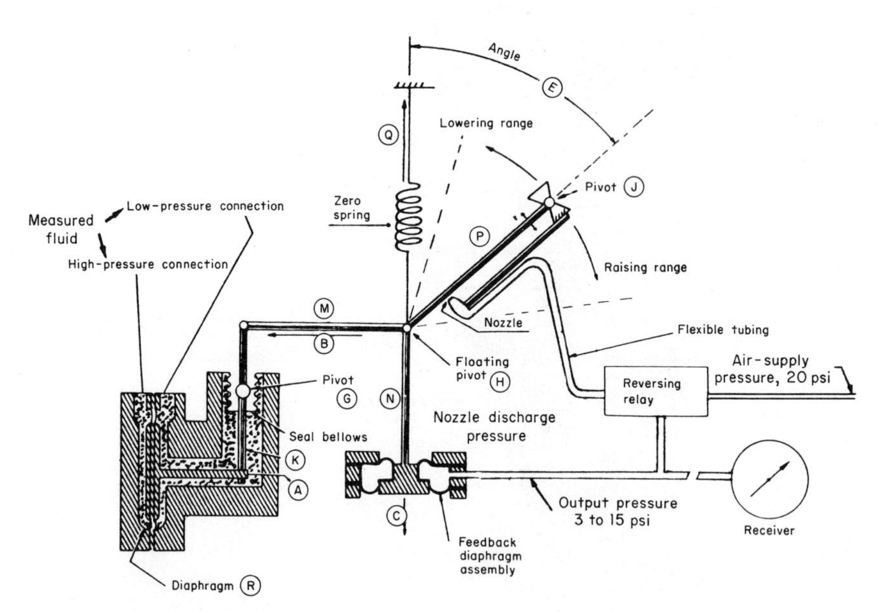

Fig. 4-4 Schematic of major components of pneumatic differential-pressure transmitter. (*Beckman Instrument Company*)

a standard output-pressure range of 3 to 15 psi is desired, tension on the vector zero spring is adjusted to produce an output pressure of 3 psi. This setting need not be changed or corrected, even if the differential pressure range of the instrument is changed.

Any desired range change within the limits of the instrument is accomplished when the position of pivot J is shifted along an arc with its center at H when in a balanced condition. This involves only loosening a screw and sliding pivot J assembly along a calibrated slot.

Fischer & Porter Company

PRINCIPLE OF DESIGN: The Fischer & Porter Company mercury manometer, of which a cross section is shown in Fig. 4-5, differs somewhat from other designs. The difference is the method used in transmitting the float motion to the pen arm or pointer. A permanent magnet located on the float extension provides a magnetic field which reacts with two magnets supported by the pen-arm drive lever to produce a magnetic coupling between the float and drive lever, thus accurately transferring any motion of the float to the pen arm or pointer, depending upon whether or not the instrument is of the indicating or recording type.

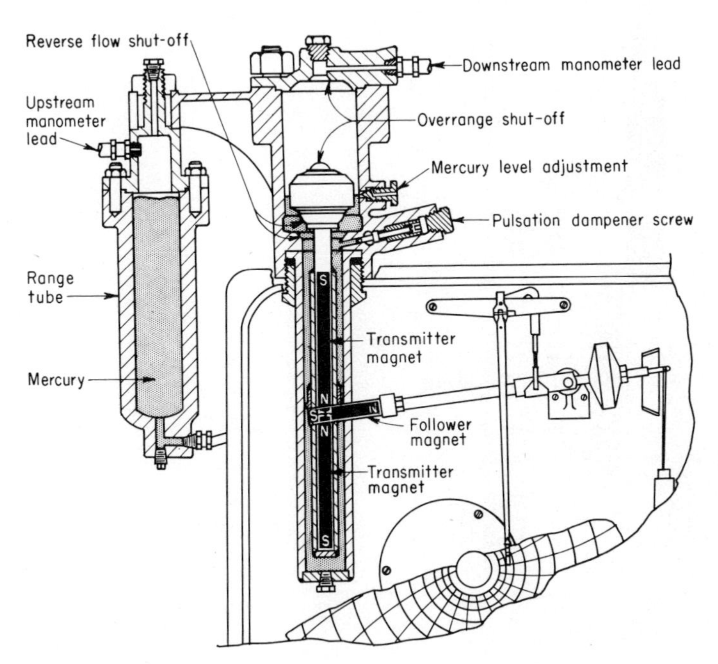

Fig. 4-5 Magnetic coupled flowmeter. (*Fischer & Porter Company*)

Fischer & Porter Company

PRINCIPLE OF DESIGN: A cutaway view of a Fischer & Porter force-balance differential-pressure transmitter is shown in Fig. 4-6. The unit consists of two metal diaphragms located on either side of a body forging, force rod, detecting pilot, air relay, force plate, and feedback bellows. The force rod extends through a Sanvik (or Elgiloy) diaphragm into a chamber which is enclosed by the two barrier diaphragms. Located within this chamber and connected to the force rod is a flexure rod which in conjunction with the Sanvik (or Elgiloy) sealing diaphragm provides a pivot around which the force rod tends to rotate when a differential pressure is applied to the barrier diaphragm.

The chamber enclosed by the measuring diaphragms and the sealing diaphragm is completely filled with stable-viscosity oil. An adjustable damping screw is located in a drilled passage between the measuring diaphragm and the barrier diaphragm on the high-pressure side. The measuring diaphragms are protected against overrange in either direction

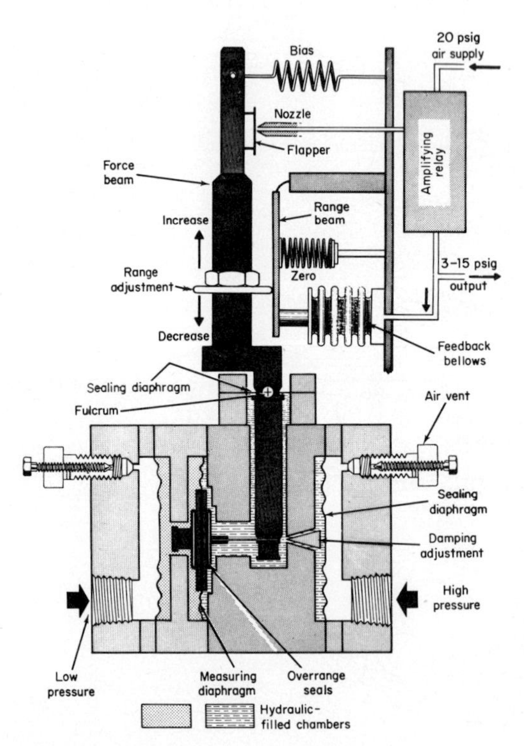

Fig. 4-6 Pneumatic force-balanced differential-pressure transmitter. (*Fischer & Porter Company*)

to a value equal to the working pressure of the unit. The temperature rating of the unit is set at —40 to +250°F.

In operation, a 20-psi air-supply pressure is applied to the air relay, and the two pressures of which the difference is to be measured are connected to the proper body connections. Any difference in pressure between the body connections results in a force being exerted on the force rod by the diaphragms, the value of which is proportional to the difference in pressure across the diaphragms. When a difference in pressure exists, the high-pressure diaphragm is forced to the left as shown in Fig. 4-6, causing a portion of the oil to flow through the adjustable damping screw into the low-pressure side of the chamber. The force exerted on the force rod tends to rotate the rod around the pivot point formed by the sealing diaphragm. Thus the extended end of the force rod which is in contact with the flapper is forced to the right, thereby increasing the back pressure in the nozzle. The increased nozzle pressure is applied to the nozzle back-pressure diaphragm of the relay, thus producing a pneumatic output proportional to the differential pressure across the primary element.

American Meter Company

PRINCIPLE OF DESIGN: The liquid-filled American Dri-Flo* meter measures differential pressure of a fluid flowing through primary devices (such as orifice plates, flow nozzles, or venturi flow tubes) for the purpose of flow measurement. Such a meter is shown schematically in Fig. 4-7. The range of such a device is a minimum of 20 and a maximum of 400 in. of water. The filling liquid of the meter bellows is a stable-viscosity liquid so that the meter's response does not change with temperature.

* Registered trade name, American Meter Company.

Fig. 4-7 Cutaway view of differential-pressure-measuring element. (*American Meter Company*)

The instrument consists of two hermetically sealed, liquid-filled bellows, joined to a common wall which divides a pressure chamber into two sections. The free ends of the two bellows are connected by a center rod. The filling liquid communicates with both bellows through a damping screw. The motion of the center rod, which is restrained by a range spring, actuates the lever arm to rotate a torque tube. A pen arm is connected by linkage to the torque-tube shaft.

If the pressure is slightly greater in one chamber (high side) than in the other (low side), the compression of the high-side bellows will cause the filling liquid to flow through the damping screw to the low-side bellows, thereby causing it to expand. The speed of response can be altered when a readily accessible external damping screw is adjusted. A small portion of the displaced filling liquid can flow through the clearance between the seal (guide washer) and the center rod as shown in Fig. 4-7. Thus, even though the damping screw is fully closed, the meter may never be entirely damped. An O-ring-seal, pulsation damper permits changing damping of the meter under full line pressure.

If a differential pressure is applied which is in excess of the range of the instrument, synthetic-rubber, flat-seal check valves which are fastened to the center rod seal and prevent any further flow of the filling liquid. The bellows are thus protected from rupturing since they are supported internally by the liquid trapped within them.

The flexible liquid-filled capsule located in the high-side chamber is connected to the main bellows system to serve as an expansion device for changes in ambient temperature. This protects the entire bellows system against damage which might otherwise result from expansion of the filling liquid and permits satisfactory operation between −40 and +170°F. When steam tracing of the gauge lines and the meter body is necessary, care should be exercised to prevent heating the meter body above 170°F. Other temperature discrepancies which might occur are compensated for by the proper use of a low coefficient of expansion materials.

All materials used which are exposed to the process fluid are of the highest-quality stainless steel or other corrosion-resistant material to ensure long life and a maximum time of trouble-free performance.

The meter is connected to the primary element in substantially the same manner as the conventional mercury-manometer-type meter. The meter can also be equipped with a pressure-measuring element and connected to the proper gauge line. Any qualified instrument mechanic can calibrate the meter because all adjustments are accessible, such as zero, range, and linearity. Instructions for calibration are included with each meter, thus giving the meter mechanic the ability to follow instructions properly.

The measuring element is protected against overrange in either direction up to the range of the meter body. The view shown in Fig. 4-7 illustrates the internal mechanism which includes a high- and low-pressure bellows, temperature-compensating capsular bellows, and torque tube. When the high-pressure bellows or a differential pressure is present across the meter, the high-pressure bellows is collapsed, and since all the cavity enclosed by the bellows is filled with liquid, part of the liquid is forced out of the high-pressure bellows into the low-pressure bellows which is resisted by the range spring connected solidly to the shaft between the high- and low-pressure bellows. The displaced liquid from the high-pressure bellows flows around the annular space of the shaft connecting the two bellows.

When the range of the meter, which is determined by the range spring, is exceeded and the check valve is set against the center plate which supports the internal mechanism, the liquid flow is sealed off, and the meter is at its maximum position. Since no additional liquid can flow, the pressure within the bellows and on the outside of the bellows is the same, and more pressure will not damage the instrument.

When the meter is overranged in reverse, the same thing occurs, except in the opposite direction. Movement of the center rod is transferred through the torque tube and the shaft which extends into the meter case where it is connected to the linkage system and the pen arm.

The meter is very accurate, rugged in construction, and simple to maintain. With proper care and a certain amount of preventive maintenance, a meter of this design should last for many years.

The Foxboro Company*

PRINCIPLE OF DESIGN: Figure 4-8 is a cutaway view of a liquid-filled, bellows-type differential-pressure-measuring instrument of the Foxboro Company design. The Foxboro meter differs in design from most liquid-filled bellows; one of the major differences is the method employed to protect the measuring bellows against overrange. Individual diaphragm disks of type 316 stainless steel, stress-relieved and heat-treated, are shaped so that they assume the position as shown in Fig. 4-8. The disks are welded around their periphery by a cycle-stitch-seam process. The diaphragm capsules are welded at the center to a concentric series of 0.080-in.-thick, stainless-steel spacing rings which butt squarely together and act as solid, metal-to-metal stops when overrange fully compresses the diaphragm.

A liquid passage is provided between the two diaphragm stacks. Within this passage is a damping valve which can be adjusted to damp

* *Ibid.*, pp. 43–44.

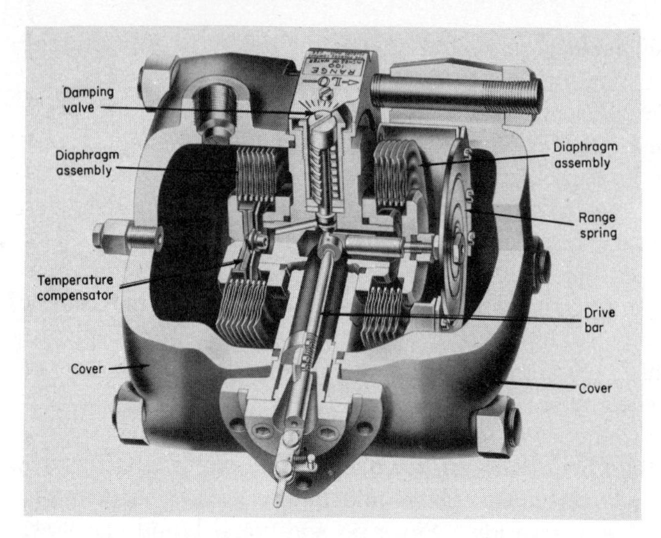

Fig. 4-8 Cutaway view of differential-pressure-measuring element. (*The Foxboro Company*)

out undesirable pulsations that might be present in the measured variable.

When the instrument is in operation, an increase in differential pressure compresses the left-hand diaphragm, as shown in Fig. 4-8. This action displaces a portion of the filling liquid into the right-hand diaphragm and expands it until the force of the range spring equals the difference between the forces acting on the two diaphragms. The linear motion of the right-hand diaphragm moves the inner end of the drive bar, and the outer end moves correspondingly through the bellows-sealed flexure. The recording pen is moved on the meter chart from the motion of the outer end of the drive bar.

Ambient-temperature compensation is provided by a bimetallic temperature element located within the compensating diaphragm, which adjusts the capacity of the diaphragm assembly to the changing volume of the fill liquid resulting from changes in ambient temperature.

The range of the instrument can be changed by replacing the range spring with one having the desired range.

The Foxboro Company

PRINCIPLE OF DESIGN: Figure 4-9 shows a cutaway view of the Foxboro model 13A differential-pressure transmitter which operates on the pneumatic force-balance principle. The measuring diaphragm assembly consists of a twin-diaphragm capsule. The capsule consists of two type 316 stainless-steel diaphragms, seam-welded to a type 316 stainless core, the diaphragms and core having convolutions. The space between the diaphragms and the core is filled completely with a temperature-stable silicon liquid. This liquid protects the twin diaphragms against over-range on either side of the capsule. The core to which the diaphragms are welded has a self-centering ring around its rim. When assembled, this core is held firmly between the high- and low-pressure castings. In this core are provided small holes so that the sealed-in fluid can pass back and forth. This feature reduces line noise to a minimum. Therefore, damping at the receiving instrument is usually not necessary. The force produced by a differential pressure across the measuring diaphragm is transmitted through a force bar to which the twin-diaphragm capsule is connected by C flexure.

A pressure seal is provided for the force bar by a metal diaphragm made from Elgiloy alloy. The top of the force bar is connected to a tension-flexure frame on which the flapper is mounted. The nozzle arrangement is securely fastened to a casting parallel to the force bar.

The pneumatic transmitting system consists of a feedback bellows, zero adjustment, air-relay valve, and nozzle. From Fig. 4-9 it can be seen that any force created by a differential pressure across the measuring diaphragms is transmitted through the force bar. This movement

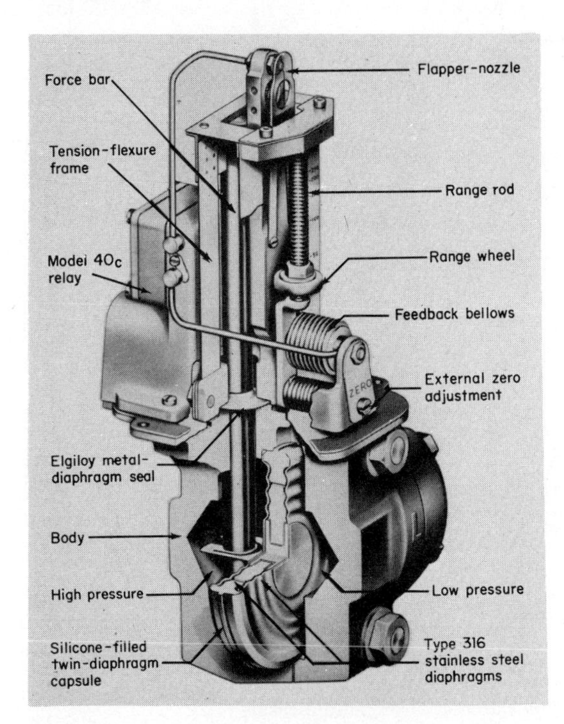

Fig. 4-9 A cutaway view of a Foxboro model 13A differential-pressure transmitter. (*The Foxboro Company*)

of the force bar operates the flapper, which in turn changes the output of the air relay in order that the pressure created in the feedback bellows will be sufficient to balance exactly the force exerted by the differential pressure across the measuring diaphragm.

Thus for every value of differential pressure applied to the measuring diaphragm within range of the instrument, there is a corresponding output signal that can be transmitted over a distance of several hundred feet and recorded or indicated as differential pressure.

Hagan/Computer Systems, Division of Westinghouse Electric Corporation

PRINCIPLE OF DESIGN: The ring-balance-type meter is designed to operate on an entirely different principle from other differential-pressure meters; the difference is in the body design. Shown in Fig. 4-10 are two drawings of a Hagan ring-balance meter.

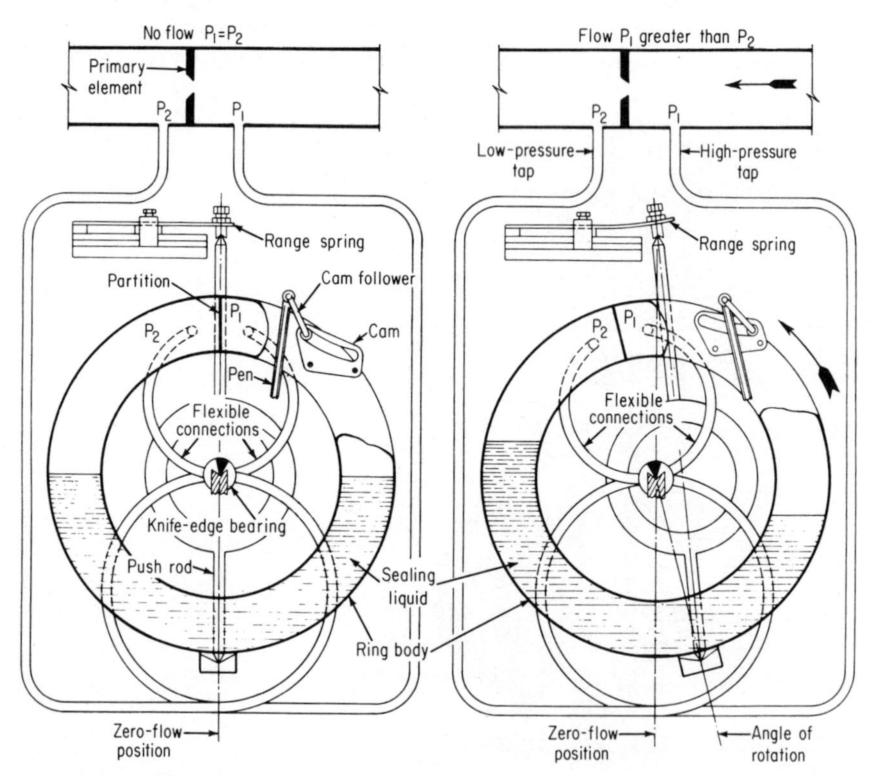

Fig. 4-10 Schematic drawing of differential-pressure-measuring element, ring-balance type. (*Hagan/Computer Systems, Division of Westinghouse Electric Corporation*)

The instrument is a radial torque meter which uses a hollow ring body to convert the differential pressure generated by a differential medium or by a difference in static pressure into a rotation which is transmitted to a recorder or indicator. The ring assembly is mounted on knife-edge bearings, which permits rotation about the axis of the ring. The ring is divided into two pressure compartments by a baffle at the top and by the sealing liquid which fills the lower part of the ring. The two ring compartments thus formed are connected to the different pressure pipes by means of flexible tubing to permit the ring to rotate freely under action of the difference in pressure in the compartments. These may be the S-shaped, self-compensating tubes shown in Fig. 4-10 for relatively high static pressure or parallel tubes for low-pressure applications.

The ring torque is a function of the differential pressure acting on the baffle. This torque is resisted by an external calibration weight rigidly attached to the bottom of the ring. The mercury or other sealing fluid exerts no force, tending to rotate the ring, but acts only as a seal for the differential pressure in the two compartments. This is true because in the circular ring body all hydrostatic forces resulting from the deflection of the sealing fluid are directed normally to the containing circle and therefore pass radially through the exact center of rotation without producing meter torque.

Standard differential pressure ranges for the ring-balance meter extend from 0.2 to 420 in. of water pressure. Such ranges are not available in any one model instrument, but the ring-balance principle is used in the various models to cover the entire range of 0.2 to 420 in. of water pressure. Various models of the ring-balance meter are designed to operate at static pressures from 5 to 10,000 psi.

The same general rule applies to the ring-balance meter as to all other recording instruments; that is, recording instruments should be installed in locations where the normal ambient temperature is below 100°F and never likely to exceed 150°F for any length of time.

Transmitting Pneumatic Instruments

The pneumatic transmitting unit shown in Fig. 4-11 is used with the Hagan ring-balance meter and consists of a sensitive pilot valve which operates to translate rotary motion of the ring into an output signal of 3 to 15 psi.

When large steam generators are used to supply varying loads of steam, cascading steam flow (or feed-water flow), boiler steam pressure, and fuel supply into the generator, the control system becomes important. In such cases pneumatic transmission of the measured variables provides a simple and effective means for bringing the signals to a centrally located control panel where they can be mixed through the use of relays to produce the desired control characteristics.

As a convenience for operations, a ring-balance pneumatic transmitter can be used to provide indication or a record at one or more points remote from the transmitter. Pneumatic receiving integrators can also be used at remote locations to provide a continuous total of flow units by attaching the integrator to the output of the transmitter.

The ring-balance-type meter is not designed for electrical transmission. However, an electrical motion-transmitting unit could be adapted to the meter if electrical transmission became absolutely necessary.

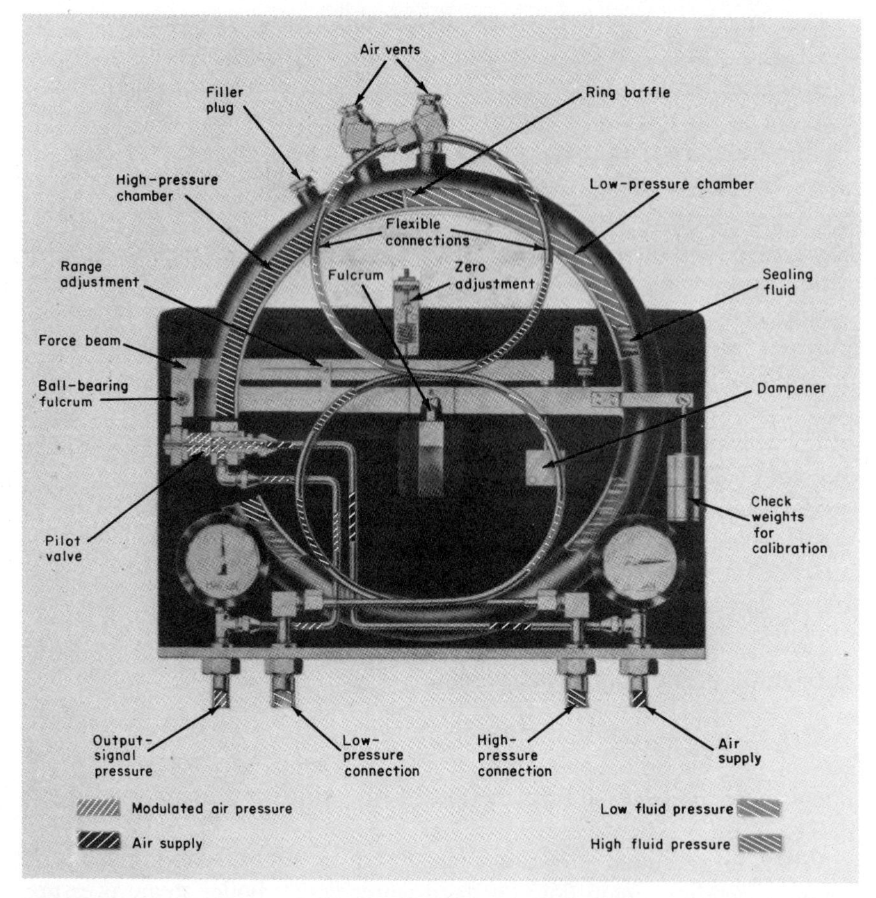

Fig. 4-11 Ring-balance-type differential-pressure transmitter. (*Hagan/Computer Systems, Division of Westinghouse Electric Corporation*)

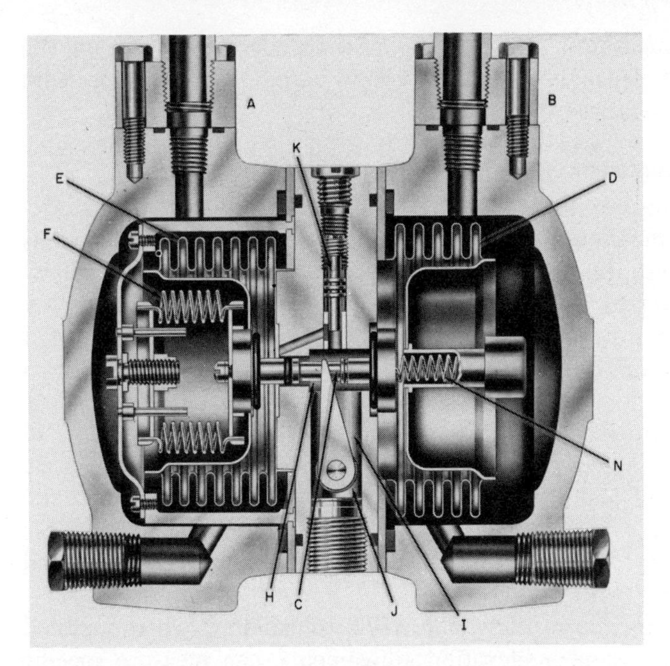

Fig. 4-12 Cutaway view of differential-pressure-measuring element. (*Honeywell, Incorporated*)

HONEYWELL, INCORPORATED*

PRINCIPLE OF DESIGN: Figure 4-12 is a liquid-filled, bellows-type, differential-pressure-measuring instrument of the Honeywell, Incorporated, design. Major components of the instrument are: high-pressure bellows *E*, low-pressure bellows *D*, torque-tube assembly *J*, range spring *F*, pressure-and-stabilizing spring *N*, and pulsation check *K*.

The two pressures of which the difference is to be measured are connected to the meter body at *A* and *B*. The higher of the two pressures is connected at *A* and the lower at *B*.

When the instrument is in operation, any difference in pressure compresses bellows *E* and forces a portion of the fill liquid through the rectangular orifice and past pulsation check *K* into low-pressure bellows *D*.

When the bellows move, bellows-connecting rod *H*, cable *C*, and torque-tube arm *I* also move. Torque-tube assembly *J* transmits this motion outside the meter body to the pointer, recording pen, or controller. Therefore, changes in differential pressure can be indicated, recorded, or used for control purposes.

Range spring *F* determines the span of differential pressure over which

* *Ibid.*, p. 48.

the instrument measures. A screw is used to adjust the zero of the meter. Spring N prevents zero shifts due to variation in static pressure and ambient temperature.

Honeywell, Incorporated

PRINCIPLE OF DESIGN: The Honeywell differential-pressure, pneumatic-type instrument shown in Figs. 4-13 and 4-14 is one of the latest designs in which some desirable features have been incorporated, such as allowing all adjustments to be made from the front without disturbing other components of the instrument.

The Honeywell $\Delta P/P$ indicating or nonindicating transmitter combines a low-volumetric-displacement meter body with a fast-responding, pneumatic, transmitting unit. Figure 4-15 is a cross-sectional view of the meter body. Differential pressures over the ranges of 0 to 5 in. to 0 to 1,000 in. of water are converted into a proportional pneumatic output of 3 to 15 psi. The output can be transmitted to any suitable pneumatic receiver for recording, indicating, or controlling the measured variable.

In addition to measuring flow, the $\Delta P/P$ transmitter, when supplied with simple suppression or elevation adjustments, can measure specific gravity or liquid level in open or closed vessels. When one pressure chamber is vented to atmosphere, the transmitter can measure process pressure or vacuum applied to the other chamber.

Overload seals, located in the center section, protect the measuring element from overloads in either direction up to the maximum 1,500-psig

Fig. 4-13 Front view of a pneumatic force-balance differential-pressure transmitter. (*Honeywell, Incorporated*)

Fig. 4-14 Front view with cover removed of a force-balanced differential-pressure transmitter. (*Honeywell, Incorporated*)

rating of the meter body (500 psi on low-range model except on special order). If the differential pressure builds up beyond normal on either side of the meter body, the appropriate seal closes. The fill behind the closed seal then increases to the overload pressure, preventing a damaging differential across either the measuring element or the dia-

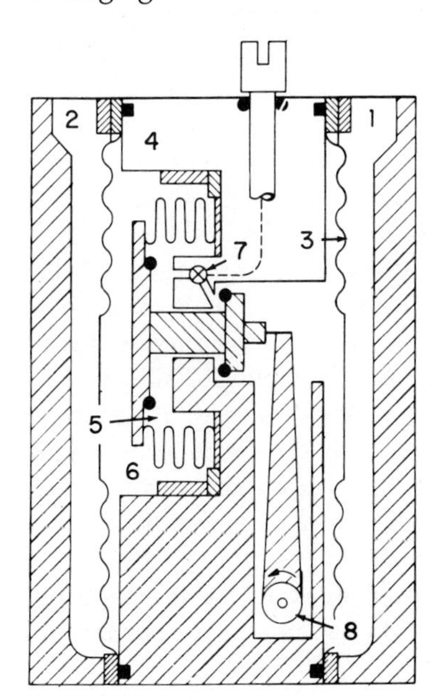

Fig. 4-15 Shown schematically is a cross-sectional view of the body of a differential-pressure transmitter. (*Honeywell, Incorporated*)

phragms. Seals are designed to open quickly when the overload subsides.

The meter body is available in almost any machinable materials. These materials are carefully chosen for long-life and high-performance characteristics.

High-speed pulsations of the process variable can be damped out for more satisfactory recording or indication. An adjustable damping restriction is located in the fill fluid of the center section where it cannot become clogged by the process fluid. This restriction cannot be completely closed; thus there is no danger that the adjustment will accidentally cause a "dead" meter.

For higher-range models the fill is a silicone fluid, inert to most materials and process fluids, noncorrosive, and a good lubricant for the internal parts; it also has the proper viscosity to give good damping as it is forced through the damping restriction. This fill fluid, when trapped by the overload seals, protects the diaphragm and measuring element if the meter is inadvertently overloaded by the application of full-process fluid pressure on one chamber only. The low-range model uses a hydrocarbon-mixture fluid fill. On applications such as oxygen service, where silicone fill is not acceptable, fluorolube fill is available.

Figure 4-14 is a front view with the front of the case removed. Note that all adjustments of the $\Delta P/P$ flow transmitter are accessible for calibration, range, and elevation or suppression.

To understand the operation of the internal mechanism, refer to Fig. 4-15 which is a cross-sectional view of the meter body. High 1 and low 2 flow-line pressures are admitted to the two end chambers. Barrier diaphragms 3 and 4 transfer these pressures, respectively, into the silicone fill fluid in the center section of the meter body on either side of the measuring element. The higher pressure acts on inside 5 and the lower pressure acts on outside 6 of the measuring element.

If there is no flow, these two pressures are equal. When the measured fluid begins to flow through the primary element, the pressure in the left chamber increases and the opposite side decreases. This unbalance in pressure causes the measuring element to move in the proper direction, forcing a portion of the filling fluid into the opposite side of the measuring element. When the measuring element moves, it exerts a proportional torque through a connecting link on the force shaft. The force shaft extends to the outside of the pressure-tight meter body through a seal tube. The transmitting unit fastens to the outer end of the force shaft.

A pneumatic balance-beam system and pilot relay in the transmitting unit converts differential pressure to proportional pneumatic signals in the range of 3 to 15 psi.

Standard tubing is oil-resistant neoprene (other tubing can be supplied). As in the meter body, transmitter operating parts are of carefully selected materials.

The transmitter is factory-set for the specified differential range. This range can easily be changed in the field by repositioning the span rider on the beam-balance assembly. A scale on the secondary beam indicates various range positions.

Zero can be adjusted either externally or internally. The external screwdriver adjustment is conveniently located on the side of the case. All models can be supplied with either elevation or suppression adjustment as a standard option.

The spring assembly can be removed or added as a field change. With elevation, the standard zero-transmitter output of 3 psi can be obtained for any applied differential from 0 to approximately 85 percent of the maximum available span, provided full-scale input does not exceed 25, 250, or 1,000 in. on the low-, intermediate-, and high-range units, respectively (e.g., an intermediate-range model adjusted for 100-in. span can have zero elevated to 150 in.). Elevation is used on liquid-level measurement of open or closed pressurized tanks with gas purging where the zero level is above the meter body (also on some specific-gravity applications).

With suppression, you can produce a transmitter output of 3 psi for an applied differential of 0 to —100 percent of maximum span up to the same limits of 25, 250, and 1,000 in. Suppression is used on all closed, pressurized, tank liquid levels (with liquid in the outer leg) and on two-way flow applications, on compound ranges, and on some specific-gravity applications.

Industrial Instrument Corporation*

PRINCIPLE OF DESIGN: Figure 4-16 is a cutaway view of a bellows-type, differential-pressure-measuring instrument designed by the Industrial Instrument Corporation. The instrument consists of five major components: the body; two measuring bellows; the shaft; and a free-floating, nonfreezing bearing. The two bellows are filled with one of several noncompressible liquids, such as glycol and water, distilled water, or special oils.

The instrument is designed with a center plate to which two bellows are attached, one on either side. The low-pressure bellows is connected to a shaft, one end of which extends through the bearing attached to one side of the center plate, where it is then connected, through a temperature-compensating bimetal U bend, to the meter shaft. The

* *Ibid.*, pp. 44–46.

meter shaft extends through the bearing assembly into the meter case where it is connected to an indicating pointer or recording pen arm.

The differential-pressure connections are made at opposite ends of the meter body, thereby allowing the two pressures to be applied to the outside of the filled system of the measuring bellows. The higher of the two pressures, of which the difference is to be measured, is connected to the right-hand end of the body (when viewed from the back).

With a zero differential pressure, the bellows are forced to the right (shown in Fig. 4-16) by the range springs connected through a bracket to the low-pressure bellows.

The meter shaft is designed with three sets of miniature ball bearings, one set of which is a combination radical and thrust bearing. The shaft seal consists of one special O ring and is permanently saturated with a special lubricant which is good for the life of the instrument. The temperature-compensating horseshoe spring, imposed between the bellows shaft and meter shaft, is designed to correct for expansion and contraction of the metal contained in the assembly, thereby producing a stable zero under varying ambient temperatures. The fluid expansion

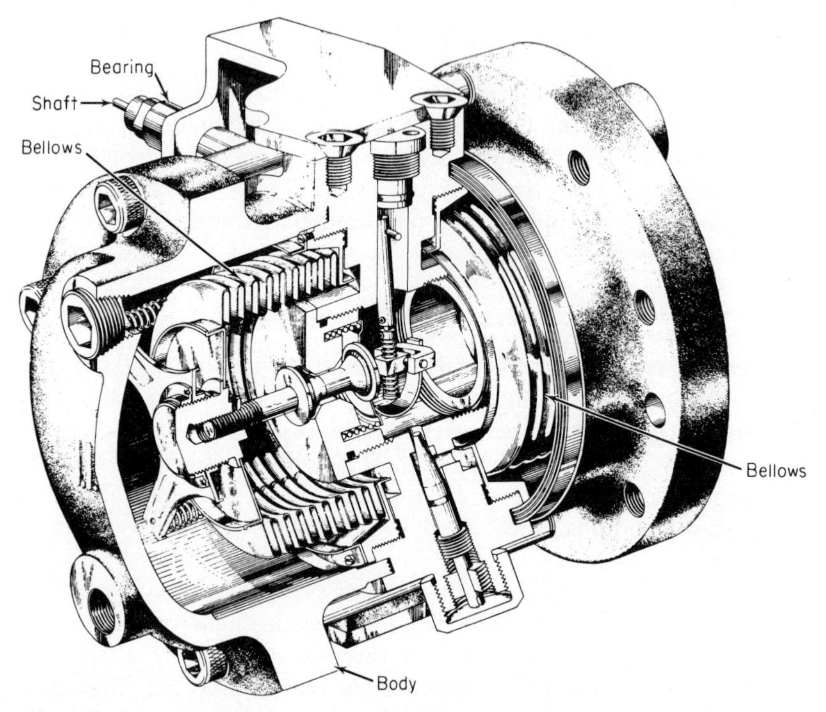

Fig. 4-16 Cutaway view of liquid-filled bellows-type differential-pressure-measuring element. (*Industrial Instrument Corporation*)

or contraction is compensated for by the bellows attached to the end of the high-pressure bellows.

Taylor Instrument Companies*

PRINCIPLE OF DESIGN: Figure 4-17 is a cross-sectional view of a pneumatic-type, differential-pressure transmitter of the Taylor Instrument Companies design. The transmitting mechanism of the Taylor model 205-T shown in Fig. 4-17 is the same as that used in the Taylor model 333-R transmitter. The measuring unit consists of eight major components: body, diaphragm, tension rod, seal bellows, high-pressure sensing element, low-pressure sensing element, and two capillary tubes.

Each sensing element and the capillary which connects it to one side of the diaphragm housing are completely filled with a stable thermal liquid. Any force applied to either of the sensing elements is transmitted by means of the filling liquid to the diaphragm housing. When a difference of pressure exists between the two sensing elements, a force proportional to this difference in pressure is applied to the internal portion of the force beam by means of the diaphragm. By action of the

* *Ibid.*, pp. 15–22.

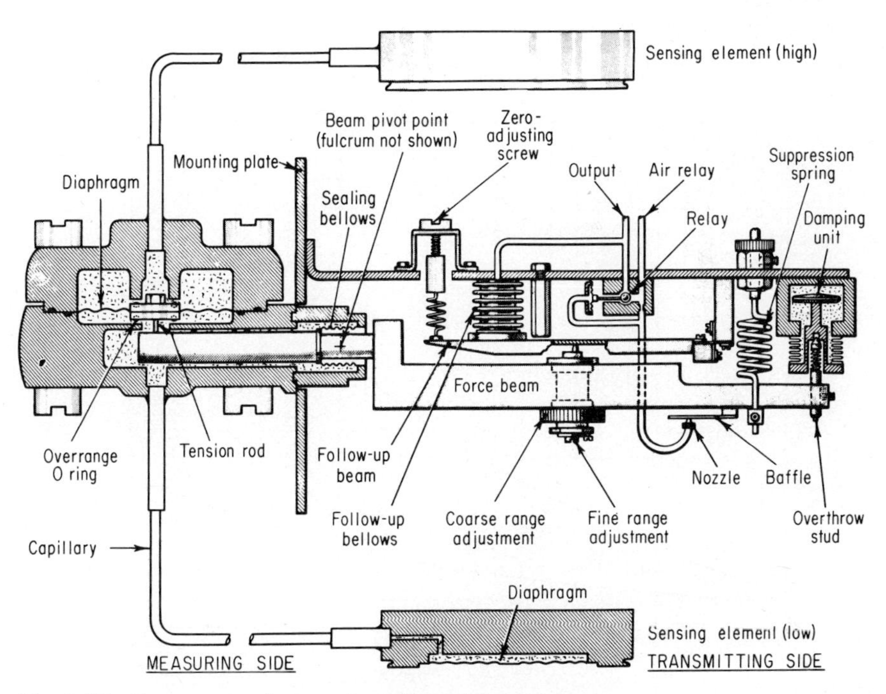

Fig. 4-17 Cross-sectional view of a differential-pressure transmitter, pneumatic type. (*Taylor Instrument Companies*)

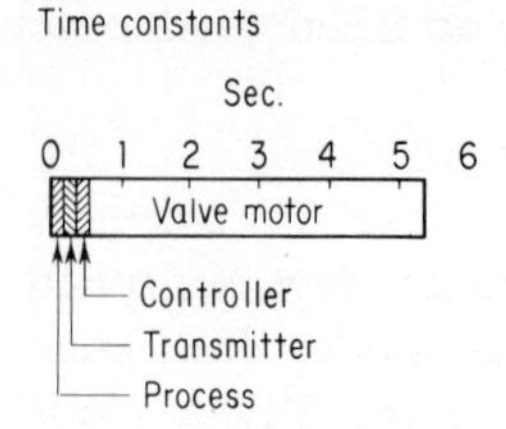

Fig. 4-18 The actual time for 63.2 percent recovery for a 3- to 15-psi step upset for flow-control system using a force-balance flow transmitter.

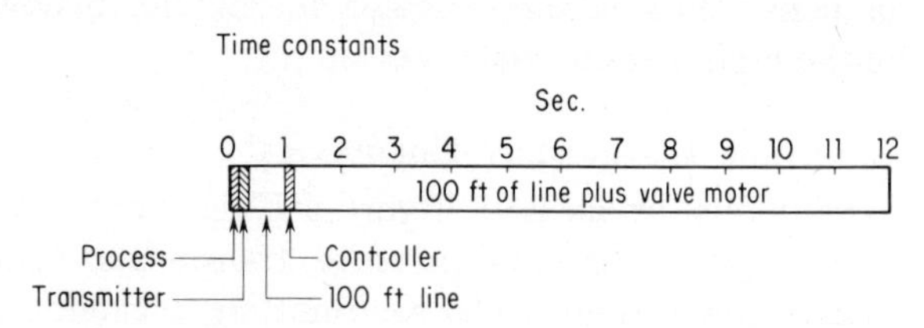

Fig. 4-19 The same as Fig. 4-18 except 200 ft of ⅜-in. tubing was used, 100 ft between transmitter and controller and 100 ft between controller and valve.

pneumatic system, consisting of baffle, nozzle, relay, and follow-up bellows, a force equal and opposite to that exerted by the diaphragm is applied to the external portion of the force beam. Thus the pressure applied to the follow-up bellows is always the exact value necessary for the force of the bellows to equal and oppose that exerted by the diaphragm resulting from a differential pressure across the diaphragm. Since the output pressure is proportional to the difference in pressure as it appears at the sensing elements, the output pressure may be indi-

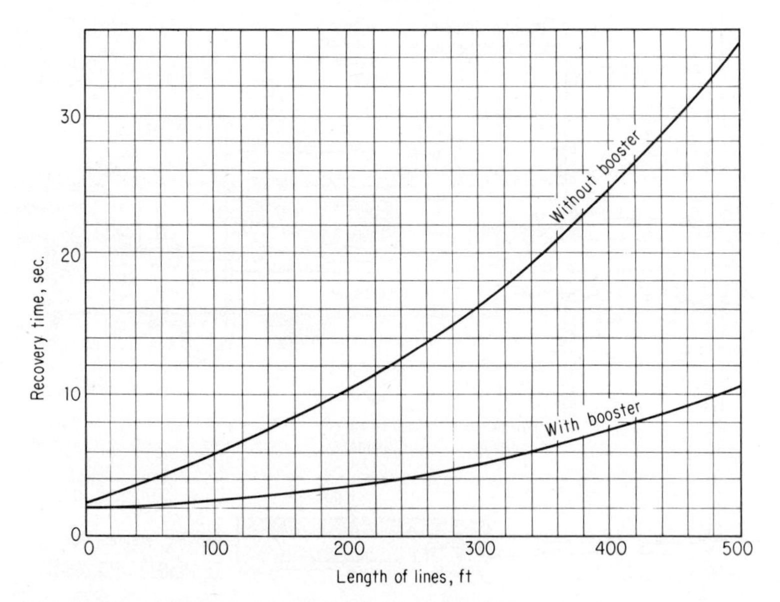

Fig. 4-20 Recovery time for various tubing lengths with and without booster relay. (*Instrument Society of America*)

cated or recorded in the desired units of measurement by a receiving instrument.

Figures 4-18 and 4-19 show the actual time for a 63.2 percent recovery for a step upset of 3 to 15 psi for the average flow-control system, using a diaphragm-type, force-balance flow transmitter. Note that Fig. 4-19 shows the recovery time with 100 ft of ⅜-in. tubing between the controller and the valve. Refer to Fig. 4-20 for a comparison of response of a control with and without a booster relay. This points out the importance of using a booster relay or valve positioner to improve the response of a control loop. However, since a valve positioner has all the advantages of the booster relay, plus the fact that a positioner eliminates the effect of control-valve-stem friction to a large degree (which is so significant for good flow control), it is preferred to the booster relay.

When the recovery time shown in Figs. 4-18 and 4-19 is compared,

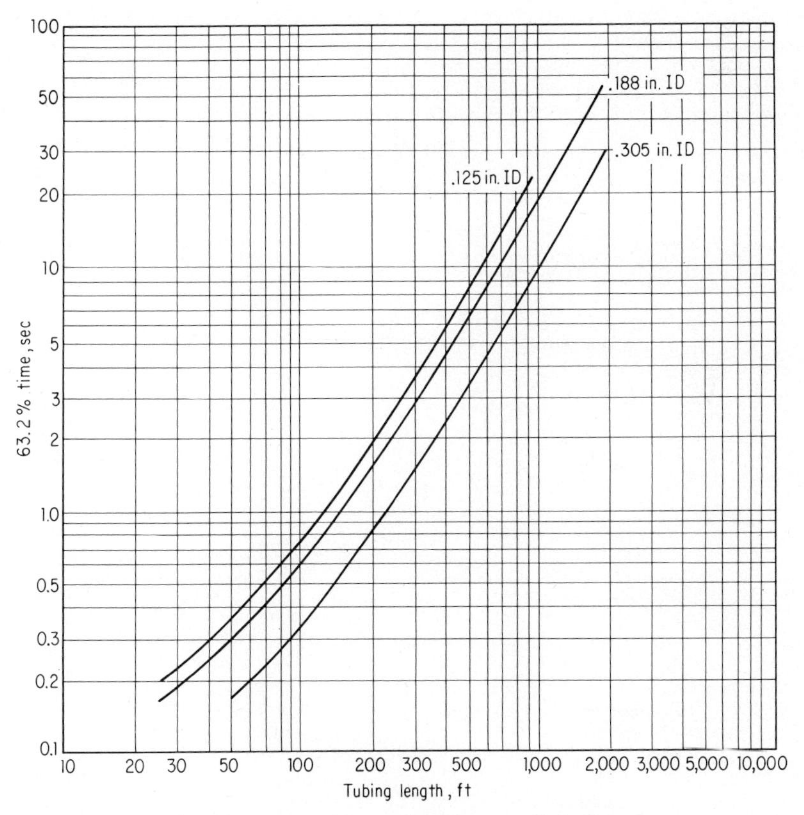

Fig. 4-21 Time constant for various tubing sizes with bellows at termination. (*Instrument Society of America*)

it is obvious that the transmission time is not the greatest offender for good flow control. The greatest difficulty is in the response of the control valve, and with the use of a valve positioner the response of the control valve can be improved to such a degree that this disadvantage becomes unimportant.

Information presented here indicates that in most cases it is practical to use diaphragm, pneumatic-type, differential-pressure transmitters for control systems where the controller is located not more than 300 ft from the transmitter and the control valve. In certain cases the distance can be extended beyond 300 ft. Time constants for various lengths and size tubes are shown in Fig. 4-21.

CHAPTER FIVE

Area-type Flowmeters

AREA-TYPE FLOWMETERS are too important to industry to treat their use lightly. Where small flows are measured the area-type meter reads directly in the units of measurement above 10 percent of its total range, which makes the meter valuable for small flows such as purge meters. These meters are designed to read larger volumes, but because their accuracy is approximately 2 percent, their use in process-control work is restricted except in certain cases.

The area-type flowmeter can be of the electrical- or pneumatic-transmission type and can be equipped with a control mechanism so that readings can be transmitted over a distance of several hundred feet.

Rotameter

The rotameter usually consists of a tapered glass tube built into a pipeline. The tube is always set in a vertical position and carries a small plumb-bob float which moves up and down as the flow increases or decreases. Graduations are etched in the side of the tube or painted on a scale to indicate the rate of flow. In some cases, small slanting notches are cut in the top flange of the bob, or float, to impart a rotating motion to it, which helps to center the float in the tube and to keep it clean of sediment that might adhere to the sides or top if it were stationary. Other designs for centering the float within the glass tube consist of ribs molded within the tube. Others may use guide rods attached to the float or those which extend through a hole drilled through the float.

Distribution of Unit-flow Graduations

One advantage of the rotameter is that the graduations do not exhibit the square-root relation encountered in the orifice type of flowmeter. It could be assumed from this fact that the area of the orifice formed by the rotameter tube and the float varies directly with the height of the float above the zero graduation. If this were so, a true linear relation would exist between flow and float position, and equal graduations would result. However, it can be easily proved that if a straight taper exists in the tube, the area of the orifice varies as the square of the height of the float. Therefore, the rotameter graduations actually exhibit a square-root relation, as is the case with orifice meter graduations.

The reason for the apparent even graduation on a rotameter scale is because the orifice area is the difference between the tube area and the float area. This permits the extension of the parabolic curve beyond the section of rapid change in slope and permits the readings to be taken on the flat section of the curve. In other words, the constants of the parabolic curve can be chosen so that they produce a curve that is nearly a straight line over most of the flow range, thus permitting the substitution of a tube with straight sidewalls. This is done by careful designing of the slope of the tube and properly establishing the ratio of the float diameter to the maximum tube diameter. If properly designed, the graduations are appreciably uneven over less than 5 percent of the full scale, and the larger graduations occur at the small-flow end of the scale and not at its large-flow end, as is the case with the orifice meter. This fact makes a rotameter a particularly advantageous meter for measuring small rates of flow.

Tapered Center-column Rotameter

If an opaque liquid is to be measured, it is possible to make the float of the rotameter visible by constructing the meter with a glass tube having a base of constant diameter and building the float in the form of a ring with an outside diameter only slightly smaller than the inside diameter of the tube. Since the ring float is always close to the glass surface, it is visible when measuring liquids that would totally obscure the conventional float in a tapered glass tube.

The variable orifice has a center column with a tapered section. As the ring float moves up the tube, the orifice size increases in exactly the same way it does with the plumb-bob float and tapered tube. It is possible to machine a curved taper on the center column that will produce exactly equal graduations throughout the range of the meter.

The Bypass Rotameter

Where large volumes of a fluid must be measured, it often becomes too expensive to build a rotameter of sufficient size to measure the total flow directly. In this case it is possible to bypass a small fraction of the flow and measure this with a small rotameter. An arrangement of this type is usually used for pipelines larger than 4 in. in diameter.

The bypass is effected by inserting a standard orifice plate in the main line and connecting the high- and low-pressure taps to a small rotameter. A secondary orifice plate having the same ratio of orifice to pipe diameter as the main-pipe orifice plate is then installed near the outlet side of the rotameter.

True proportionality would exist between the flow through the main and the bypass pipes if the friction loss in the bypass were the same as that in the main pipe and if the pressure drop were concentrated at the orifice in the bypass. This ideal situation cannot be achieved in practice, but it is approached by designing the rotameter float as light as possible to reduce the pressure loss at this point. The pipe friction is reduced as far as possible by using a low ratio of d/D, thus reducing flow through the bypass and consequently the pipe friction.

All the problems of orifice location that were found to affect the measurement of flow by means of orifice plates are applicable to bypass rotameter installations. Since the installation has a standard orifice as the primary element, it is possible to calibrate the rotameter against a simple mercury manometer by arranging pressure outlets on the bypass line. Readings can then be taken on the rotameter and manometer in sequence. This calibration verifies the degree of proportionality achieved on the particular installation and permits corrections to be made to the flow readings where necessary.

Piston-type Area Meter

The piston-type area meter, a second kind of meter, operates in the same manner as a rotameter. The principle of design of this meter is similar to that of the rotameter as seen in Fig. 5-1, although its general appearance is entirely different. In this instrument the flow enters the meter body horizontally, is deflected against the floating piston, and passes through an orifice on the side opposite the inlet. The orifice is rectangular in shape and fixed on the bottom and sides. The top of the orifice area is formed by the piston, and as the latter moves up or down, the area of the orifice is varied accordingly. A bypass leads from the low-pressure side of the meter to the top of the piston so that the pressure on the top of the piston is always the same as that on the downstream side of the orifice.

In operation the piston moves up and opens the orifice just far enough to maintain a constant-pressure differential between the high- and low-pressure sides of the orifice. This is exactly the operation of the rotameter float, and the position of the piston is therefore an indication of the rate of flow of the liquid passing through the meter.

In this case the movement of the piston has a linear relation to the orifice area since only one dimension of the orifice varies. This being the case, the flow graduations are equally spaced. Note that in the case of the rotameter, the orifice area varies as the square of the diameter, and hence the relationship between orifice area and float height is quadratic.

The range of the piston is varied by changing its weight (which may be as heavy as 5 lb) and by varying the width of the orifice. Four different widths of orifice are available.

The recording mechanism for this meter is identical with that of the electrically recording flowmeter. An armature extends from the piston into two induction coils which are a part of a bridge circuit. Any movement of the piston unbalances the bridge circuit, which is brought

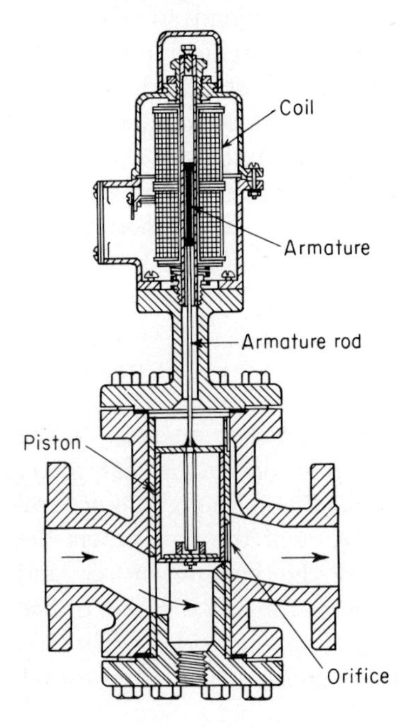

Fig. 5-1 Piston-type area meter. (*Honeywell, Incorporated*)

to balance by a corresponding movement of the pen-actuating armature in the recorder branch of the circuit.

The meter can be supplied with a steam-jacketed body if it is necessary to maintain the fluid at a high temperature in order to prevent sticking of the liquid in the piston cylinder. It is also built with a parabolic-shaped orifice if desired, which changes the flow-chart graduations from linear to the familiar square-root form. This is sometimes desired where the meter is in a control system adjusted to control on the basis of uncompensated orifice-meter readings.

Proper installation requires that there be at least 20 pipe diameters of straight pipe in front of the meter. Note that this is a requirement not needed in the case of the rotameter. It is also desirable to provide a bypass around the meter in order to permit removal of the meter without shutting down the pressure.

Variable-area-type Element

Transmitting, Pneumatic

Brooks Instrument Company*

PRINCIPLE OF DESIGN: Shown schematically in Fig. 5-2 is a pneumatic transmitter designed by the Brooks Instrument Company, Inc., which is used in conjunction with various types of Brooks rotameters. The instrument is a magnetically actuated, indicating mechanism combined with a pneumatic transmitter whose purpose is to provide a flow-rate indication and an air-output signal which is linear with the flow rate as measured by the rotameter on which the transmitter is mounted.

The indicator consists of a pointer which rotates through an arc of approximately 85° over a horizontal indicating scale. A magnet is mounted on the end of the float extension (see the schematic diagram). The magnetic position converter, which is mounted close to the float-extension housing, consists of the flat strip of steel twisted into a helix. The edge of the helix turns as the magnet in the float moves up and down. Because of the construction of the magnetic position converter, the magnetic lock is unbreakable, and the pointer of the magnetic position converter reproduces the float position exactly.

The pneumatic transmitter is activated by the flow-sensing cam mounted on the magnetic-position-converter drum. The cam is characterized in accordance with the fundamental calibration data of the rotameter. A pair of nozzles acts as a cam follower. Air at approximately 6 lb is supplied to the lower nozzle and blasted against the

* Grady C. Carroll, *Industrial Process Measuring Instruments*, 1st ed., McGraw-Hill Book Company, New York, 1960, pp. 80–81.

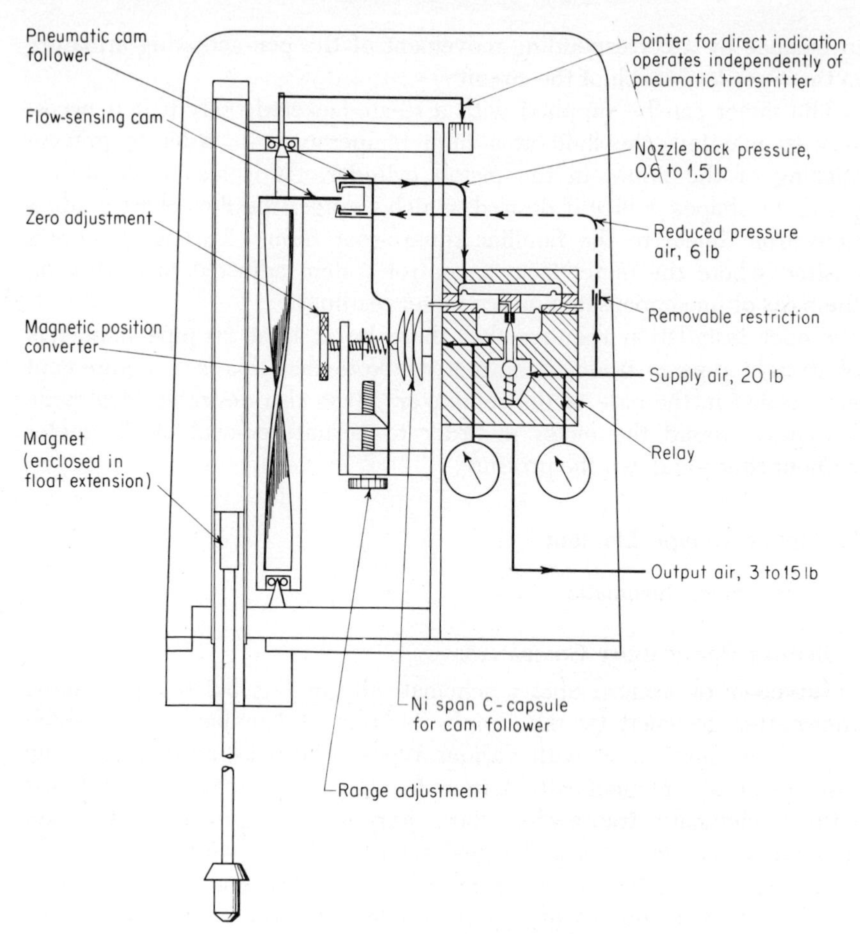

Fig. 5-2 Pneumatic transmitter. (*Brooks Instrument Company*)

upper nozzle, which is connected to one chamber of the relay, shown in the schematic diagram as the top chamber—actually the relay is mounted so that this is the front chamber. When the flow rate increases the magnet contained in the float extension rises, which causes the magnetic position converter to rotate, and the cam is withdrawn from between the nozzles. This action permits the air blast from the lower nozzle to be directed more fully against the upper nozzle with resulting higher pressure on the diaphragm of the upper relay chamber. This forces the diaphragm assembly down, opening the relay valve and resulting in higher output pressure. The output pressure is directed into a follow-up capsule, which repositions the nozzles closer to the edge of the cam. When flow rate decreases, the magnet drops, and the cam is rotated more fully, blocking the air passage between the nozzles and thus reversing the relay and follow-up-capsule action. This system produces an air output depending upon the contour of the cam, which is accurately contoured at the factory to match the flow-rate conditions existing. Zero and span adjustments are provided. No linearity adjustment is required because linearity is built into the cam.

FISCHER & PORTER COMPANY

PRINCIPLE OF DESIGN: Figure 5-3 is a Fischer & Porter Company variable-area differential-pressure meter which bears the trade name of V/A Cell.* When a flowing liquid in a pipeline is measured, a primary device such as an orifice plate is installed between a tapped flange, as it would be for a mercury-manometer installation. The V/A Cell is connected to the taps on each side of the orifice, thus providing a bypass through the V/A Cell. Any differential pressure across the orifice

* Registered trade name, Fischer & Porter Company.

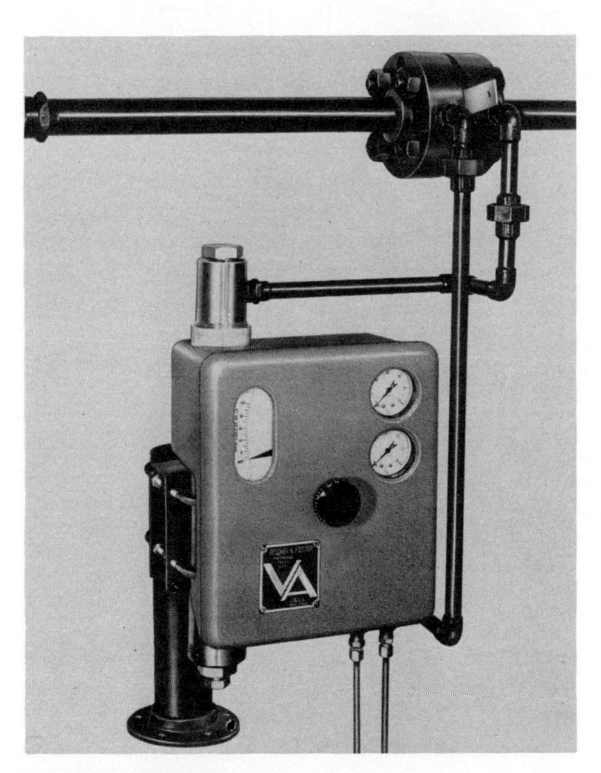

Fig. 5-3 Differential-pressure area meter—V/A Cell. (*Fischer & Porter Company*)

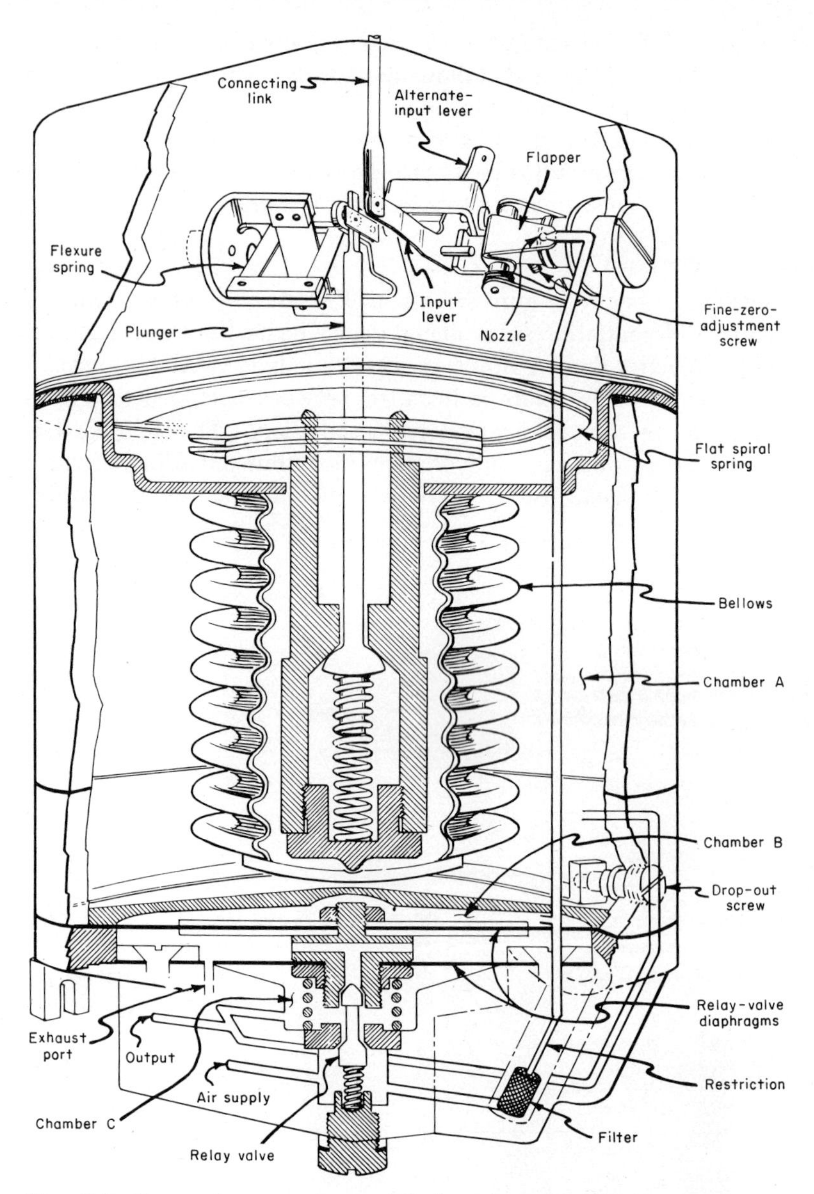

Fig. 5-4 Pneumatic transmitter. (*Fischer & Porter Company*)

causes a fluid flow through the V/A Cell, thereby producing a reading on the scale which is directly proportional to the volume of fluid flow through the orifice.

The "kinetic manometer," as Fischer & Porter Company refers to the V/A Cell, has several highly desirable features. The two which will appear to the application engineer most are its uniform scale and elimination of packing glands. To eliminate packing glands, Fischer & Porter Company uses a magnetic bond between the float extension and the indicating arm. There are two permanent magnets fastened to the float extension with their north poles together. The indicating arm is designed to operate coupled at this point, and any tendency of the arm to depart from it is repelled by the two south poles of the two magnets attached to the arm, thereby making it impossible for the indicating arm to operate at any other point on the magnet or to become uncoupled from the magnet if its bearings are free.

The Fischer & Porter V/A Cell body can be manufactured in almost any material which can be machined. Its maximum operating pressure is 1,000 psi at 100°F. The differential pressure range extends from 0 to 50 to 0 to 400 in. of water pressure.

The Fischer & Porter V/A Cell shown in Fig. 5-3 is designed to use the standard Fischer & Porter transmitting unit shown in Fig. 5-4 which consists of nozzle, baffle, relay valve, and follow-up bellows. Motion of the float is transmitted through the magnetic coupling and pointer arm to the nozzle and baffle where other components of the system operate to convert the arm motion into a pneumatic output signal of 3 to 15 psi.

The linear characteristic of the V/A Cell relative to volume results in a pneumatic output signal that is also linear to the volume of fluid flowing through the primary element to which the cell is connected. This one item is the greatest asset of the cell and gives it an advantage in some cases over differential-pressure transmitters whose pneumatic output signal is the square of the flowing factor as related to volume of a fluid flowing through a primary element such as an orifice, flow nozzle, or venturi tube to which the transmitter is connected.

In large processes where continuous blending of solids and fluids is necessary, signals of all measured variables can be made linear to volume by the use of orifices or flow nozzles with a V/A Cell for the fluid measurement. A simple linear-cascade-control system is then possible.

CHAPTER SIX

Volumetric Flowmeters

A LARGE AND IMPORTANT GROUP OF FLOWMETERS is classed as the volumetric type. The volumetric meter is fundamentally different from the inferential meters that have been considered in the previous sections on flowmeters.

Volumetric flowmeters definitely have a place in modern industry—for filling and emptying tanks, filling barges and tankers, transferring material from one plant to another, as well as measuring small quantities around homes and in the laboratory. They also serve for accounting purposes and measuring various materials in complex installations. Volumetric flowmeters are extensively used in chemical-process and allied industries in suitable metering applications.

Piston Type

The simplest volumetric flowmeter that has the largest number of variations in design is the piston-and-valve type. Almost any kind of piston pump will (if pressure is applied to the fluid in the discharge side) operate as a motor. If the number of reciprocations of the piston or the number of rotations of the crankshaft is counted, the volume of fluid passing through the system can be computed, provided the dimensions of the cylinder and length of stroke are known. Furthermore, the volumetric efficiency of a pump operating as a meter may be considered as 100 percent since the piston chamber must be completely filled to produce the power to drive the mechanism. Designing a piston-type flowmeter resolves itself into the construction of a reverse-acting pump that has the proper characteristics.

A somewhat different design of piston meter is constructed with a vertical reciprocating piston operated by means of a nutating disk mounted below the cylinders. The disk is pivoted in the center, and each piston is connected to it by connecting rods mounted in ball-and-socket joints built into the disk. The nutations of the disk thus operate the pistons in sequence and also rotate a valve mechanism that permits each cylinder to be filled and emptied in succession. This meter is an adaptation of the simple nutating-disk meter, and will be more clearly understood after study of this kind of meter.

The piston-type meter is particularly well suited to the measurement of liquids under low head pressure and at low flow rates. The lowest rate of flow that can be accurately measured is dependent upon the leakage rate. This leakage factor becomes increasingly significant as the flow rate decreases, and in all piston-type meters a flow rate is established below which the measurements are out of commercial tolerance. However, with a meter that has tight pistons and is in good working condition, the volumetric efficiency is nearly 100 percent, and the meter will measure very close to zero flow.

The accuracy of this type of meter varies from 0.2 to 0.3 percent, depending upon the condition of the fluid. The effect of liquid density is negligible, and any change in viscosity of the liquid being measured also produces a negligible error. The pressure loss through the meter is in large part dependent upon the restriction on the valve parts. This being the case, the pressure differential required to operate the meter will vary as the square of the flow rate. This is true only when the piston areas are large enough to reduce to a negligible value the effect of the friction forces in the meter.

The piston-type meter operates satisfactorily on all fluids at all temperatures, provided the fluids are sufficiently clean to prevent wear on the cylinder walls, pistons, and valve seats and are not too viscous to flow freely or are not chemically corrosive. A strainer should always be used to trap foreign material before it can reach the meter body. Where dry liquids are being measured, such as gasoline or naphtha (which have little or no lubricating value), it is usually necessary to use a greater initial pressure head on the meter to prevent sticking because of lack of lubrication in the moving parts. In this case the pistons are usually sealed with leather cup washers since this design produces less friction than a piston-ring seal.

The capacity of the piston-type meter ranges from measurements as low as $\frac{1}{20}$ gal per min in the 1-in. size to as high as 300 gal per min in the 3-in. size. The cost of meters of this type in larger than 3-in. sizes is prohibitive for most installations. In the smaller sizes the piston meter operates on only a few inches of head pressure. The maximum static working pressure for meters of this type is usually 150 lb per sq in.

Nutating-disk type

The nutating-disk-type flowmeter, shown in Fig. 6-1, is one of the oldest and most widely used because it is a successful instrument for domestic water-metering. The principle of the meter is simple, and there is but one moving part.

The moving part consists of a nutating disk pivoted in the center of a chamber which has hemispherical-shaped sides and a cone bottom. The inlet is on one side of the meter and the outlet diametrically opposite. The disk has a vertical, centrally located shaft which fits into the pivot socket in the center of the meter and rotates about a cone on the top side—the rotation about the cone drives a gear train which actuates the counting mechanism. The differential pressure of the liquid causes the disk to nutate about the center pivot, and since the fit between the disk and the hemispherical sides is close, the nutations are a direct measure of the volume of liquid passing through the meter. The meter is therefore a true volumetric meter.

The nutating-disk meter is usually made of cast iron with a carbon or brass disk, the former being used for service where great sensitivity is required. This is because the carbon, being light in weight, responds to lower rates of flow. However, the carbon disk is essentially brittle and should not be used where the meter is subject to shock or waterhammer. If corrosive liquids are to be measured, the meter body and working parts may be made of corrosion-resisting materials such as

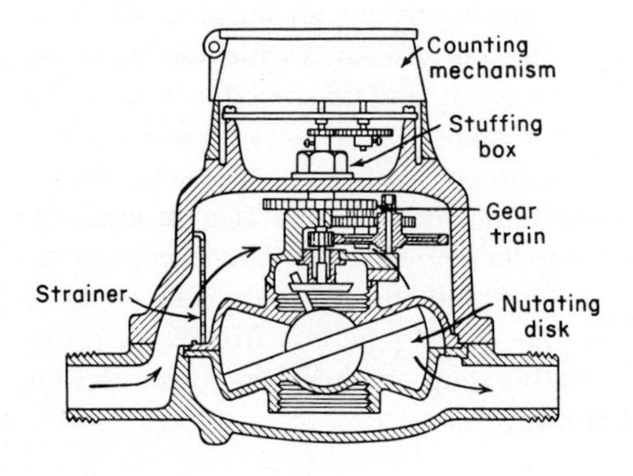

Fig. 6-1 Nutating-disk meter.

bronze, brass, aluminum, or Monel metal, depending upon the nature of the corrosive material. The operating pressure of this type of meter may be any pressure up to 1,000 lb per sq in., and the meter is particularly adapted to the measurement of liquids under high pressures and temperatures.

The meter is usually built in sizes up to 4 in. and accurately measures flows from 1 gal per min in the ½-in. size to 375 gal per min in the 4-in. size. As in the case of the piston type, sizes larger than 4 in. are too costly for most applications.

The accuracy of the nutating-disk meter is not so high as that of the piston type because it is necessary to allow a greater clearance between the disk and the meter walls than is required between the piston and cylinder in the piston meter. However, the liquid does not have to be so clean and free from abrasive material to work satisfactorily.

Calibration is effected by changing the ratio of the gear train or by means of a needle-valve adjustment which permits corrections in flow to compensate for wear or change in viscosity or specific gravity.

Since the nutating-disk meter is not completely sealed, a certain amount of liquid is bypassed, and for this reason the meter is not impervious to errors due to viscosity effects. At high rates of flow the meter shows an error of about 1 percent because of a change in viscosity from 40 to 700 in saybolt universal units.* If the meter is calibrated for a liquid of medium viscosity, the error resulting from viscosity changes does not usually throw it out of commerical tolerance.

The effect of density on this type of meter is negligible in most cases, although it is not so insensible to changes in density as the piston-type meter, which is entirely free from this source of error. The pressure loss across the meter varies as the square of the flow, as with the piston meter.

Volume-measuring Elements

Propeller-type Element

B I F, Unit of General Signal Corporation†

PRINCIPLE OF DESIGN: The meter shown in Fig. 6-2 is a product of B I F, Unit of General Signal Corporation. It is a propeller-type, totaliz-

* Everett M. Cloran, "Rotary Oil Meters of the Positive Displacement and Current Type," *Trans. ASME*, vol. 60, November 1938, p. 617.

† Grady C. Carroll, *Industrial Process Measuring Instruments*, 1st ed., McGraw-Hill Book Company, New York, 1960, pp. 107–109.

ing flowmeter with a direct-reading, six-digit totalizer. The high-tensile, cast-iron body has a venturi shape to increase velocity in the region of the propeller, thus increasing sensitivity and accuracy at low flow rates. Full-section straightening vanes are built into meters of all sizes. Because of the venturi-body design, straightening vanes, and streamlined internal structures, the B I F meter (Propeloflo*) operates with minimum disturbance and minimum loss of head. The propeller is made of molded plastic of high-impact strength. Internal gear and shaft housings are cast bronze. Gears are wear-resistant phosphor bronze, and shafts and ball bearings are stainless steel. Metals other than copper and copper alloys are available as special equipment.

In operation, the liquid entering the venturi-shaped body is directed by the straightening vanes to the propeller which rotates at a speed proportional to the rate of flow of the liquid passing through the meter. This rotation of the propeller is transmitted through bevel gears and spur gears to the totalizer mechanism, where suitable gearing actuates the six-digit totalizer, providing flow totalization in the proper volumetric units. The totalizer may also be geared to read in weight units at a specified constant density.

Table 6-1 shows the available ranges, pressure, and temperature limitations for various sizes of Propeloflo meters.

* Registered trade name, B I F, Unit of General Signal Corporation.

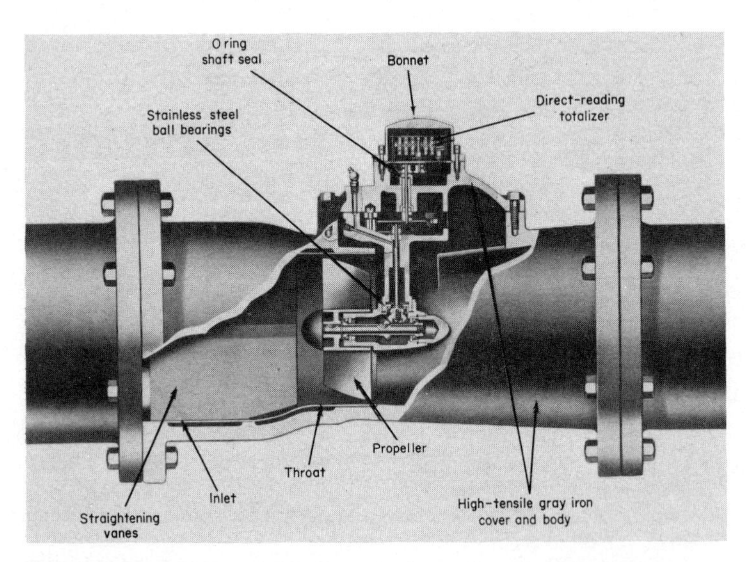

Fig. 6-2 Propeller-type flowmeter. *(B I F, Unit of General Signal Corporation)*

TABLE 6-1 Ranges, Pressure Limitations, and Temperature Limitations of B I F, Unit of General Signal Corporation Propeloflo Meter

Size, in.	Max. flow, gpm	Min. flow, gpm	Max. oper. press., psi	Max. oper. temp., °F
2	90	25		
3	215	35		
4	400	40	300	225
5	600	60		
6	900	85		
8	1,300	90		
10	1,800	110	300	225
12	2,400	140		

SOURCE: Grady C. Carroll, *Industrial Process Measuring Instruments*, 1st ed., McGraw-Hill Book Company, New York, 1960, p. 109.

Turbine-type Element

B I F, UNIT OF GENERAL SIGNAL CORPORATION*

PRINCIPLE OF DESIGN: Figure 6-3 is a cutaway view of a turbine-type flowmeter, a product of B I F, Unit of General Signal Corporation, called Shuntflo.† The meter is designed for in-line installations in 1- through 4-in. sizes. For lines larger than 4 in., a 2-in. meter is used in a bypass around an orifice plate mounted between flanges in the main line as shown in Fig. 6-4.

The meter shown in Fig. 6-3 consists of a high-tensile-iron or cast-steel body, high-tensile-iron or cast-steel cover, and a combined cooling-and-damping chamber. Within this chamber are the working parts of the meter, consisting of the fan-shaft assembly, reduction gear-train assembly, and alnico driving magnet. The main body also contains the orifice and nozzles which are sized to give the desired meter range. The fan-shaft assembly is supported by and rotates on a jewel bearing. The damping fan near the bottom of the shaft is submerged in the damping liquid which normally fills the cooling-and-damping chamber. The blades of the damping fan are pitched at an angle to produce a downward thrust to balance the upward thrust of the driving fan, thereby considerably reducing the axial loading in the bearings.

Below the cooling-and-damping chamber is the counter box, which contains a counter, alnico following magnet, and the necessary gearing

* Carroll, *Industrial Process Measuring Instruments*, pp. 109–115.

† Registered trade name, B I F, Unit of General Signal Corporation.

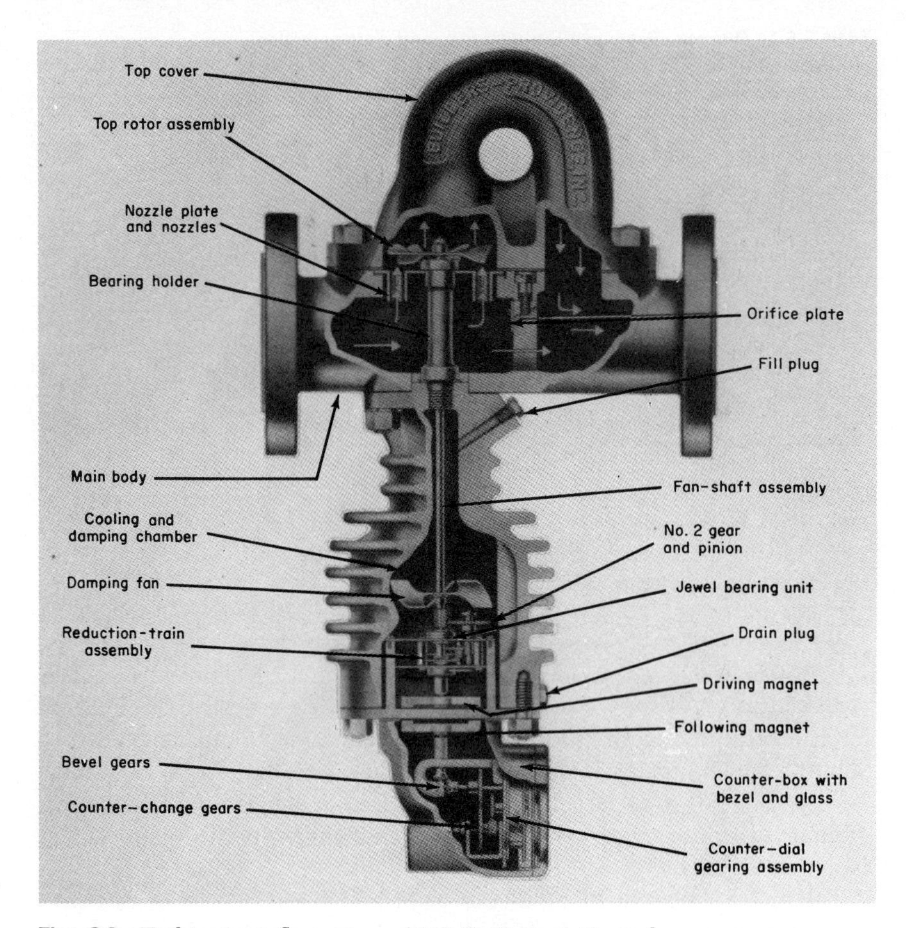

Fig. 6-3 Turbine-type flowmeter. (*B I F, Unit of General Signal Corporation*)

to cause the counter to register in the desired units of flow. To eliminate a stuffing box, which would require a packing gland or pressure seal (a potential source of trouble), a magnetic coupling is provided between the drive magnet in the damping chamber and the following magnet in the counter box to drive the counter mechanism. Pressure-compensated counters are now available which automatically correct the counter reading for changes in line pressure for compressible fluids (air, gases, steam).

Steam, air, or gas to be measured flows through the meter body with a portion passing upward through the nozzle, impinging on and turning the driving fan. The portion passing through the fan flows through the shunt passage in the top cover and rejoins the main flow, passing on through the meter outlet. The speed of the fan shaft is dampened and the thrust minimized by action of the damping fan which rotates in liquid contained in the damping chamber. The portion of flow which acts upon the drive fan is proportional to the total flow at all rates within the normal range of the meter. Therefore, the speed of the fan shaft is proportional to the volume of fluid flowing through the meter. In bypass meters, as shown in Fig. 6-4, the orifice in the Shuntflo body is blank except for a small drain hole. When used in a bypass installation, most of the flow is through the mainline orifice.

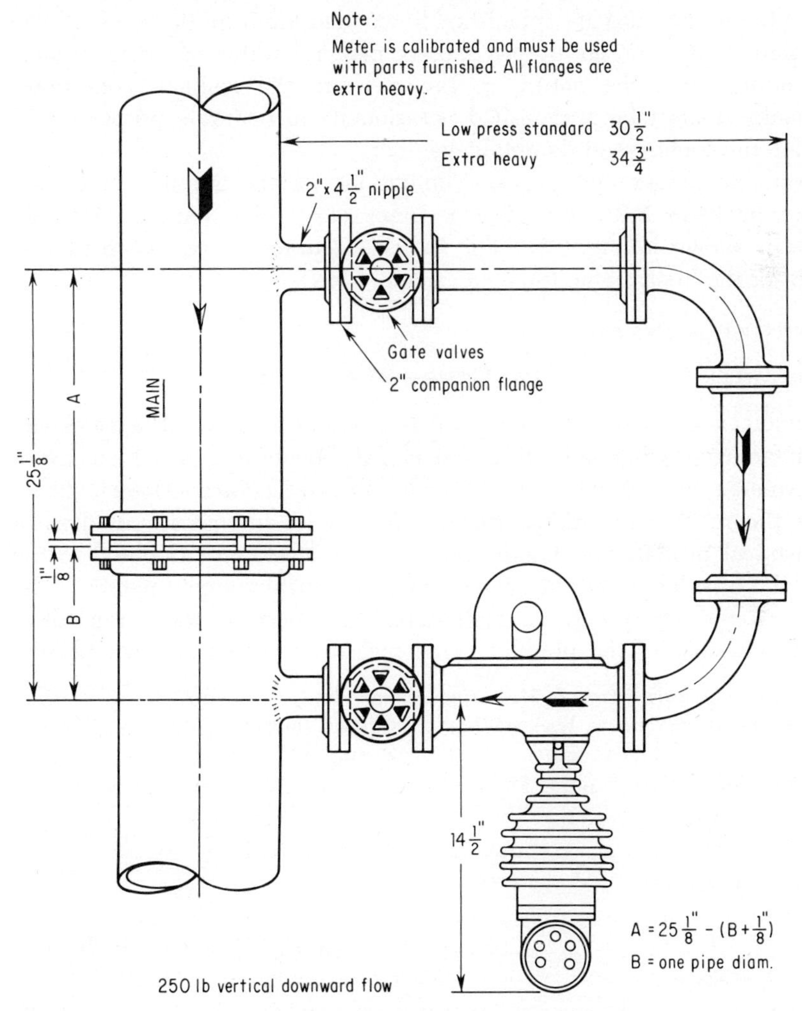

Fig. 6-4 Shuntflo meter. (*B I F, Unit of General Signal Corporation*)

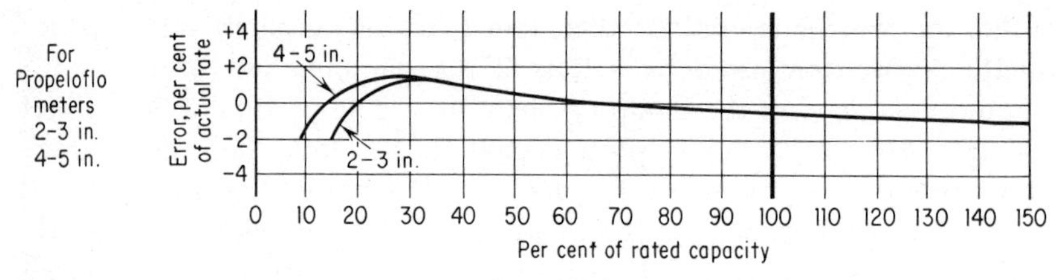

Fig. 6-5 Typical accuracy curves for Shuntflo meter. (*B I F, Unit of General Signal Corporation*)

Damping liquid for steam meters is water which, after the initial filling before first use, is replenished by condensate from the steam. The damping liquid for air or gas meters may be water or other liquids, depending upon the nature of the gas and the ambient conditions. Normally it must be replenished occasionally to maintain proper liquid level in the cooling-and-damping chamber.

Available ranges and pressure limitations for the Shuntflo meter are shown in Table 6-2. A typical accuracy curve of a standard Shuntflo meter is shown in Fig. 6-5. The typical accuracy curve shown in Fig. 6-6 is for the bypass Shuntflo meter model SMKS.

Turbine-type Element

Rockwell Manufacturing Company*

PRINCIPLE OF DESIGN: Figure 6-7 is a cutaway view of a Rockwell Manufacturing Company 16-in., series "M" meter designed to measure the volume of a flowing liquid. The Rockwell Turbo-Meter† differs from the more conventional turbine-type meter in that it can drive a mechanical upshaft and handle on-meter registration as opposed to the typical ac voltage output signal on most turbine-type meters. The Turbo-Meter incorporates a system for sustained accuracy, regardless of the viscosity of the product being measured, which is accomplished

* Carroll, *Industrial Process Measuring Instruments,* pp. 103–107.
† Registered trade name, Rockwell Manufacturing Company.

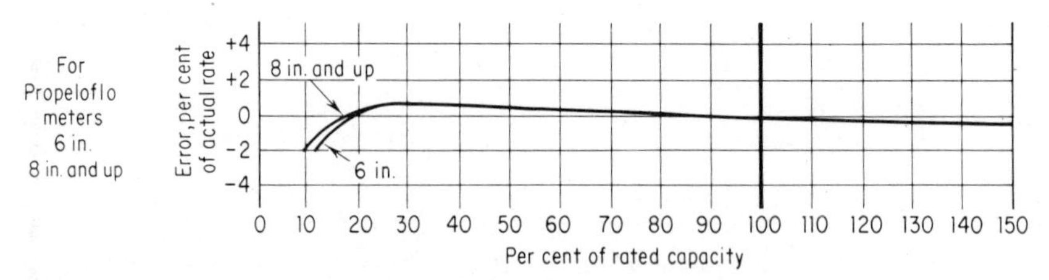

Fig. 6-6 Typical accuracy curves for bypass Shuntflo meter model SMKS. (*B I F, Unit of General Signal Corporation*)

TABLE 6-2 Ranges and Pressure Limitations of B I F, Unit of General Signal Corporation Shuntflo Meters*

Meter size, in.	Model	Flow range	Max saturated steam capacity lb/hr, at stated pressure, psig					Max air capacity cfm, at stated pressure, psig			
			10	50	100	150	200	10	50	100	150
								Wide variety of meter nozzles and orifices available to obtain any desired capacity within limits shown			
1	SMKS	†	86	268	353	420	477	27.2	88	118	142
2	SMKS	*	1,190	3,150	4,150	4,940	5,620	447	1,122	1,502	1,803
3	SMKS	*	2,500	7,180	9,450	11,250	12,790	867	2,445	3,275	3,930
4	SMKS	*	4,430	13,200	17,390	20,700	23,520	1,475	4,420	5,920	7,100
5	SMKS	†	4,850	14,600	19,230	22,890	26,020	1,540	4,800	6,380	7,650
6	SMKS	‡	6,870	20,950	27,580	32,830	37,320	2,190	6,880	9,150	10,960
8	SMKS	‡	11,960	36,860	48,530	57,770	65,670	3,800	12,100	16,100	19,300
10	SMKS	‡	18,500	57,320	75,470	89,830	102,130	5,880	18,800	25,000	29,950
12	SMKS	‡	26,490	82,330	108,400	129,030	146,690	8,450	27,000	36,000	43,100

* 10:1. Approximate rated capacities in lb per hr (steam) and cu ft per min (air). Air capacities are in units of free air at 14.7 psig, 60°F, and 60 percent saturated, assuming air in line at 60°F and 60 percent saturated. For line temperature other than 60°F, multiply by $\sqrt{520/460 + T}$. For gases other than air, divide by $\sqrt{\text{sp gr of the gas}}$. Air and gas meters should never be operated above rated capacity.

† From 4:1 to 10:1, depending upon pressure and maximum rated capacity.

‡ 7:1.

SOURCE: Grady C. Carroll, *Industrial Process Measuring Instruments*, 1st ed., McGraw-Hill Book Company, New York, 1960, p. 114.

by the inclusion of an integral-viscosity compensator for which the Rockwell Manufacturing Company has been issued a patent. Since the basic turbine-meter principle depends upon the relationship between the rate of product flow and the rotor speed, any change in viscosity affects the meter accuracy. The viscosity compensator in the Turbo-Meter provides and maintains the required constant relationship between product flow rate and rotor speed, thus improving the meter accuracy.

Shown in Fig. 6-8 are typical accuracy curves for a Rockwell 16-in, series "M" meter. Figure 6-9 shows typical head-loss curves for a Rockwell 16-in., series "M" meter.

In the simplified sketch in Fig. 6-10, the actual movement of the flowing product can be readily followed. The flow shown by the broken arrows enters the meter and passes around the upstream diffuser, through the rotor (causing it to rotate), around the downstream diffuser, and out of the meter. Housed within the upstream diffuser is the viscosity-compensator case and drum, the latter directly coupled to the

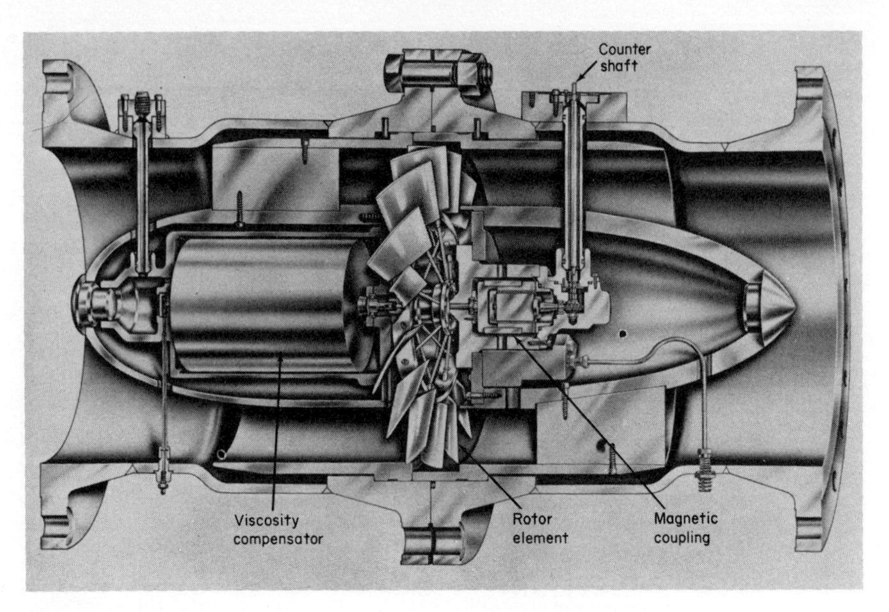

Fig. 6-7 Turbine meter 16-in., series "M." (*Rockwell Manufacturing Company*)

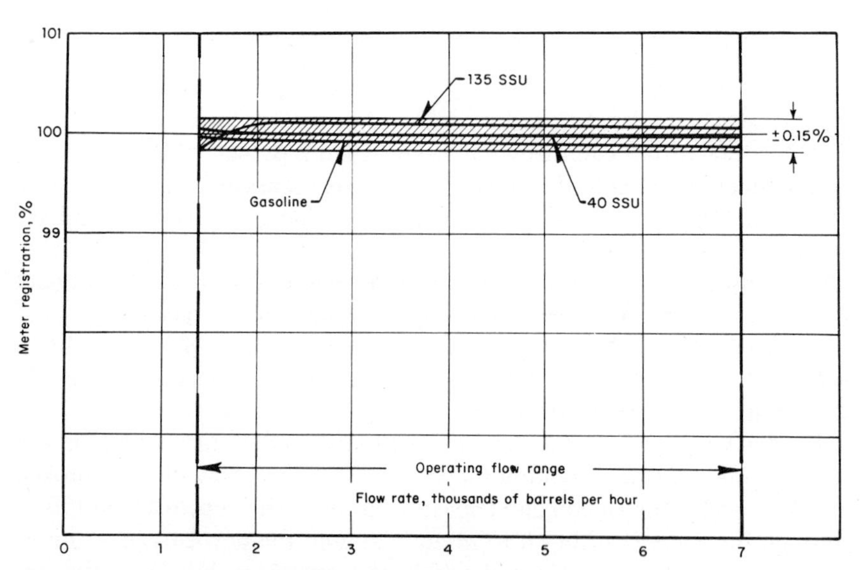

Fig. 6-8 Typical accuracy curves for a Rockwell 16-in., series "M" meter. (*Rockwell Manufacturing Company*)

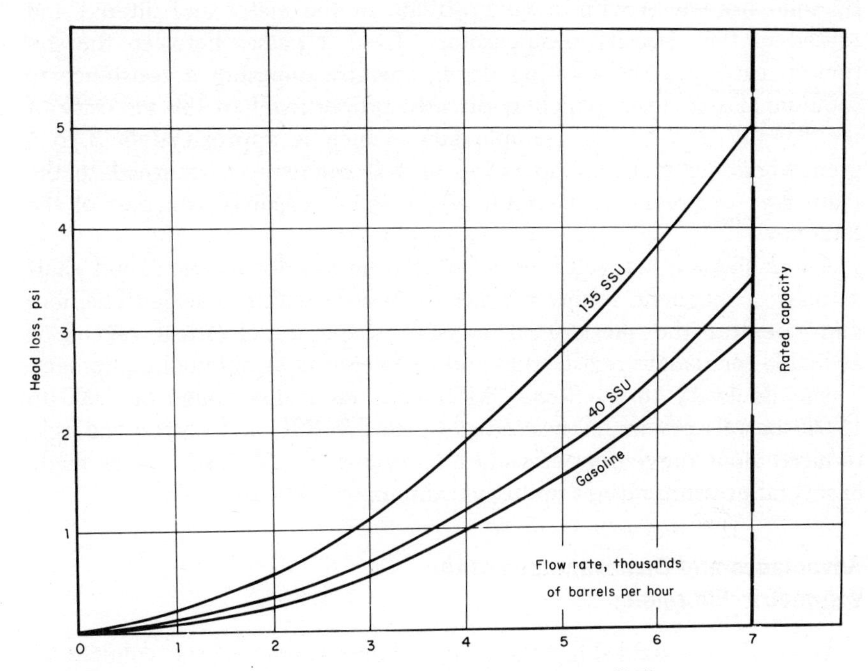

Fig. 6-9 Typical head loss curves for a 16-in., series "M" Turbo-Meter.

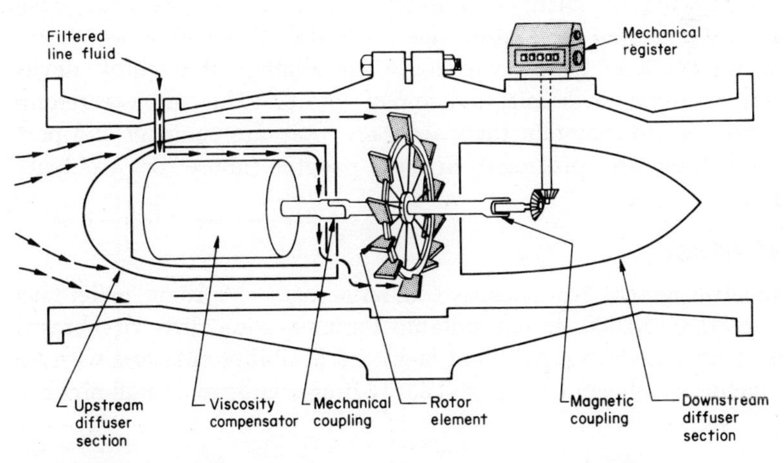

Fig. 6-10 Simplified sketch of Turbo-Meter. (*Rockwell Manufacturing Company*)

rotor and rotating with it. A small portion of the flowing liquid, shown by solid arrows, is withdrawn upstream of the meter and filtered and routed to the viscosity compensator. Here it passes between the stationary case and the rotating drum, thereby imposing a resistance to rotation of the drum which is directly proportional to the viscosity of the flowing liquid. The sample flow, which is approximately 3 to 5 gpm when the meter is operating at full capacity, is returned to the main flow upstream of the rotor where it is measured as a part of the total flow.

A fork-shaped, magnetic driver located on the downstream rotor shaft actuates a magnetic follower which is housed within a sealed chamber and operates the mechanical register. Separate electrical takeoff is available for remote registration and telemetering as optional equipment.

The Rockwell 16-in., Series "M" meter has a flow range of 3,000 to 15,000 barrels per hr for viscosities up to 275 SSU and correspondingly reduced flow range for viscosities in excess of 275 SSU. It is made in several pressure ratings up to a maximum of 1,440 psi.

Advantages and Disadvantages of the Volumetric Flowmeter

As an aid in establishing the range of applications of the volumetric-type flowmeter, an analysis of its advantages and disadvantages in chemical-process work is enumerated.

Advantages

The volumetric meter is accurate on small rates of flow and is not seriously affected by changes in pressure, density, or viscosity. Such meters have a simple integrator, and the installation cost is low. Volumetric meters lend themselves to the installation of a cutoff device to permit metering a definite amount of liquid. They do not require level conditions to maintain their accuracy, they are compact and portable, and they are protected against possible abuse or accidental breakage.

Disadvantages

Volumetric meters are expensive in large sizes. A 10-in. meter may cost more than $3,500, is not suitable for measuring dirty or abrasive liquids, is expensive to repair, and lacks the flexibility obtained with the orifice meter, which can be changed easily from a large to a small pipeline.

CHAPTER SEVEN

Automatic-control Theory

AUTOMATIC CONTROL can be defined as the mechanical technique of integrating the responses obtained from primary measuring instruments, discussed in previous chapters, and producing counterresponses calculated to maintain a state of internal balance in a process under control. To those concerned with process work in the chemical, metallurgical, or allied industries, any mechanical means capable of maintaining steady states of internal balance in a process system is of enormous value. The continuous chemical process is so dependent upon the instruments which control it that it would be necessary, if these instruments were not available, to confine the operations of chemical manufacturing to a simple unit process hardly capable of producing the cheap chemical goods available today. Listing the advantages of automatic-process control would be the same as listing the continuous-chemical process itself and will not be dealt with here.

Definition of Heat-exchanger Lags

All the aforementioned factors are significant from the standpoint of control only because they have a definite affect on the lags that are inherent in the exchanger. The first lag shown at the top of Fig. 7-1 is titled the "demand-side storage lag" and represents that portion of the total volume of the heat exchanger filled with the medium whose temperature is to be controlled. The word "lag" infers a time element that is not evident in the foregoing description. To make this concept clear, let us assume a sudden change in the demand for the material going through the heat exchanger. The time required for a certain definite temperature deviation to be detected at the outlet of the heat exchanger depends upon the capacity of the demand side of the heat exchanger. This can be explained by the analogy between the capacity of an electric iron and an electric light. When in operation both of these appliances may reach a state of dynamic heat balance; when they are both suddenly turned off, the light filament loses its heat almost instantly, whereas the iron remains hot for some time. This difference in the time required to reach a new temperature equilibrium is due to the difference in heat-storage capacity of the two systems. In this case the iron has a much greater storage lag.

Types of Controller Mechanisms for Coping with Process and Response Lags

The preceding paragraphs clearly indicate the complexity of the problem of automatic-process control. No less than seven sources of lag have been analyzed as existing in the control circuit of a process. Other lags can also be introduced in a particular process because of an unusual

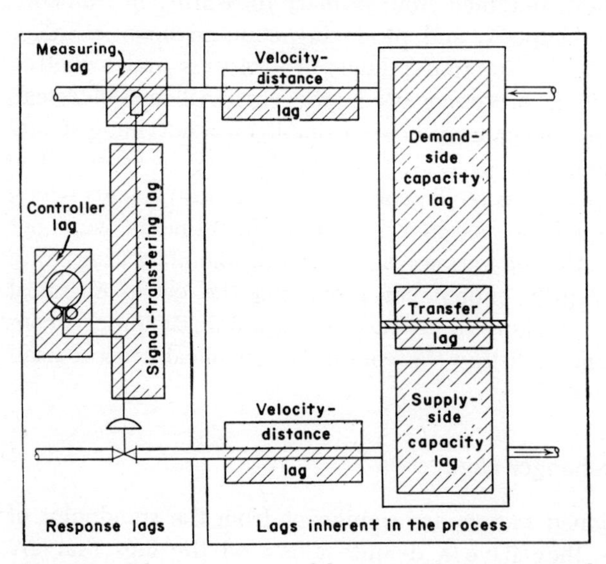

Fig. 7-1 Inherent sources of lags in temperature-control system.

condition not existing in a standard process. The only proper way to consider a control problem is to study it in its entirety. Discussion of a type of controller without reference to the exact application for which it is intended has little value since its effectiveness depends entirely upon whether it is suitable for the particular process to be controlled. Likewise, any mathematical approach to the integrated problem of control must state clearly the exact process conditions for which the analysis is valid. If this is not done, an apparent simplification of the analysis results which may be misleading.

All the sources of lag analyzed so far are inherent in the equipment making up the controlled system; and although they may not all exist in any particular process, several are sure to be present. Good design will hold these lags to certain desirable proportions but cannot eliminate them. Once the process is designed, the problem remains of choosing a controller mechanism with characteristics that assure the minimum specified deviation from the control point necessary for successful operation of the process at the least cost for control equipment. The more complex and unfavorable the lags are in process, the more elaborate and expensive the controller mechanism must be to cope with them. The fact should not be overlooked, however, that good control design is measured by the simplicity of the control mechanism. This simplicity represents lower first cost for equipment and lower maintenance costs. Too often elaborate controls are installed on a process where the demands on the system never require the use of the more complex response for which the controller was built.

If we assume that a process has been designed properly, it is now necessary to choose a control mechanism that will function in such a way that it will minimize the effects of the lags that are present. There is a variety of ways in which the controller may respond, and these different modes of control are now listed with a short description of the characteristic action of the controller. The actual characteristic of the response that may be expected from each type will be considered in a subsequent section. The controller mechanism is of great importance in the problem of control since it is the only means available for counteracting the inherent instability in a continuous process. Those available are as follows:

1. Off-and-on control
2. Proportional control
   a. Narrow band
   b. Wide band
3. Floating control
4. Proportional-and-floating control
5. Proportional-and-floating control with valve positioner

Off-and-on Control

The off-and-on control is the simplest and most often used type of control mechanism and is clearly described by its title. Control of this type is also known as "two-position control" because the valve can assume only two positions regardless of the temperature of the process. The two positions usually assumed are the open and shut positions of the valve, although valves are made that can be adjusted for other than these positions.

The valve may be electrically, compressed-air, or hydraulically operated, or operated directly by the temperature-measuring fluid. In all cases, however, the characteristic of the valve is such that it moves rapidly and positively from one position to the other, depending upon whether the temperature of the process is higher or lower than a certain prescribed point. When the valve is open, the heating rate is too high; when it is closed, the rate of heating is too low. To prevent the large disturbance that may result from having the supply shut off completely by the control valve, a bypass valve is often devised to admit a steady supply of a minimum amount of heat known to be needed and to have the off-and-on valve supply the additional variable amount required for control. This method may produce difficulties, as will be discussed in a later section.

An extension of this simple off-and-on type of control results when a third position for the valve is added. In this case the valve normally stays in a central position and moves to a fully open or fully closed position, depending upon the direction of temperature deviation. This is similar to the bypassed type of open-and-shut control, except that it may have a more elaborate control mechanism to provide the three positions.

Proportional Control

The first refinement of the simple basic off-and-on control was the construction of a control mechanism that would cause the control valve to assume a position proportional to the deviation of the temperature from the point of control. In this type of controller there is one position of the controlling valve for each temperature within the temperature range encompassed by the open and shut positions of the valve. This temperature range is known as the "control band" and is illustrated in Fig. 7-2.

The width of the band varies with the type of controller mechanism, of which there are two classes designated as: (1) narrow-band controllers and (2) wide-band controllers.

The range of the control band with respect to the total range of the temperature-measuring device of the controller is usually represented as a percentage or ratio. Under these classifications the narrow-band type of controller usually has a range of approximately 10 percent of the full scale of the controller. The wide-band type may have a throttling range equal to 100 percent of the temperature range of the instrument or may be equal to several hundred percent of this range. In the latter case the valve is open at all times and moves but a small amount for a normal-temperature deviation. The term "throttling range" is very common in the language of control and describes the temperature band shown in Fig. 7-2. This is a descriptive term because the proportional-control valve is exactly described as a "throttling" valve.

Most of the proportional-control mechanisms have an adjustment that permits the width of the control band to be varied over the entire throttling range. The set point about which the control operates is also variable throughout the entire temperature range of the instrument. Where only a narrow band is required for control, the narrow-band-controller mechanism is used since it is less costly. Where doubt exists as to the workability of this type, the wide-band controller is used because it can be adjusted to operate satisfactorily in a narrow band if the process requires this range. There is no functional difference in these two types of controllers, although we shall see later that a change in mechanism is needed to achieve the wide-band characteristics.

Floating Control

A third control mechanism, described as a "floating" control, produces a valve movement that is independent of the amount of deviation of the temperature from the control point. When the temperature is above the control point, the valve closes at a constant rate and continues to close until the temperature drops below the control point. At this point the valve reverses its action and moves in the other direction until the control point has been passed again. This type of control is essentially a cycling control and is used only where the valve movement can be made to slow down considerably.

A modification of this type of control is to arrange a two-speed opera-

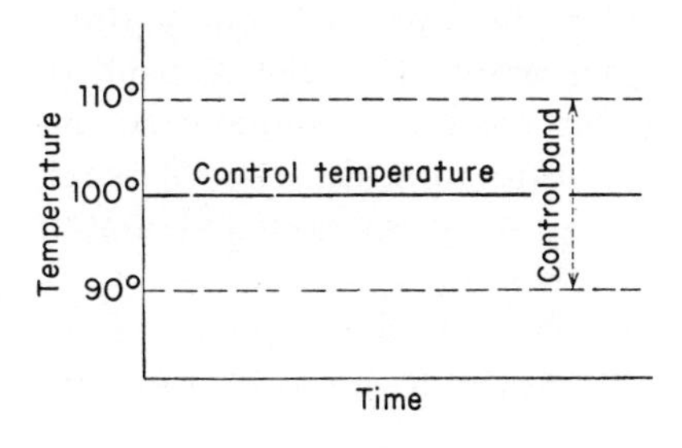

Fig. 7-2 Width of control band.

tion which provides for a slow movement of the valve over a narrow portion of a control band and develops a higher valve speed if the temperature reaches beyond these limits in either direction. This action improves the control by permitting slow valve travel within the desired temperature range and gives a quicker correction if a large deviation should occur.

A second modification of the floating control similar to that in the previous paragraph is to provide a dead zone, or temperature band, in which the valve makes no move. In this case the valve only "floats" after a certain set deviation has occurred, which helps to reduce the cycling that would otherwise result from the slightest disturbance.

The fourth type of floating control is more complex and is known as the "proportional-speed floating" control. Here the valve moves constantly in one direction or the other, depending upon whether the temperature is above or below the control point, but moves at a rate that is proportional to the deviation of the temperature from the controlled point. This control may also move at a rate that is some other function of the deviation of the temperature than normal. However, the movement is usually at a rate in direct proportion to the temperature deviation.

This type of control is of considerable interest, not because it is used often as a mode of control but because it forms the basic movement from which is derived the next type of control mechanism which is very useful and widely used.

All these floating controls work best where the lag in the system is small and where the response is rapid. In this control even the demand-side storage lag is not desirable because it promotes cycling.

A further use for the floating control is to control the position of the temperature-control point where an additional variable in the process must be taken into account. This is done, however, only in unusual installations where the greatest accuracy of control must be obtained.

Proportional-and-floating Control

A further refinement in the control mechanism results from a combination of the proportional-position control (just described) and the floating control. In the proportional-position control there was only one position of the valve for each temperature existing within the control band of the operation. We shall see later that this condition produces an undesirable deviation of the process temperature from the control point, and to correct for this, the control mechanism is so arranged that it reacts simultaneously to the impulses of the two forces.

The first is the floating-control force, which is the dominate force producing a movement of the valve. It moves to position the valve exactly as it would in the case of floating control just described. Superimposed on this movement is a second proportional-control force which tends to increase the valve opening while the temperature is decreasing and to decrease it while the temperature is increasing. This second corrective action has no permanent effect on the valve position, which is determined entirely by the floating control.

In order to enable the controller to cope with variations in the process, adjustments are provided for varying the strength of these two corrective forces separately. Sometimes a single adjustment is also provided which varies both these adjustments simultaneously so that, when one force is increased, the other is automatically decreased a proportional amount.

Proportional-and-floating Control with Valve Positioner

With the use of the more complex type of control mechanism, any variable lag in the controller caused by friction or any lack of positive power to operate the controlled valve seriously disturbs the action of the controller. Friction in the controlled valve produces a hysteresis action which prevents the valve from moving until a large enough force is applied to the valve mechanism to overcome this friction resistance. The result is an additional lag in the process which may cause trouble because of its size and because it will be unpredictable and variable, depending upon the tightness of the valve packing, the oil in the mechanism, the temperature, etc.

To overcome this variable, an additional mechanism is provided that permits the application of enough additional power on the valve mechanism to overcome this friction without changing the ultimate position of the valve, which is determined entirely by the amount of the corrective action applied through the means described in previous paragraphs. Usually this additional control mechanism is incorporated as an intimate part of the proportional-and-floating-control mechanism, as will be discussed in the section covering the mechanical details of these instruments.

All the types of control mechanisms that have been described can be studied methematically by means of a schematic diagram which has all the parts needed to produce the valve movements considered. Each type of response can then be studied by causing secondary reactions not needed in the simpler controllers to become inoperative. This analysis has been made by Spitglass* and will be considered here in detail, using the nomenclature of the original study.

Figure 7-3 represents a typical controller capable of reproducing any

* Albert F. Spitglass, "Quantitative Analysis of Single-capacity Processes," *Trans. ASME*, Pro-60-9, November 1939, pp. 665–675.

Fig. 7-3 Temperature controller. (*Beckman Instrument Company*)

of the responses mentioned in the previous sections. The initial process-signal movement is applied to the sensing element of the controller. This is transmitted by a lever to a pilot valve of the controller which adjusts the output to the control valve until the process returns to its desired position.

CHAPTER EIGHT

Process Characteristics Using Different Control Mechanisms

EACH OF THE CONTROL MECHANISMS described in the previous chapter has a different effect on the standard continuous process when used as the controlling element. Knowledge of this resulting control is the crux of the entire control problem, and these characteristics must be thoroughly understood if an intelligent selection of equipment is to be made. All the preceding study has been preliminary work to acquaint the reader with the nature of the process to be controlled and the mechanical means available to achieve this control. Little has been said of the characteristics of the control resulting from the application of a particular control mechanism in the process.

Control Mechanisms

Off-and-on Control

Graphic representation of a simple open-and-shut control is given in Fig. 8-1. The interrelated variables involved are the demand, the controlled temperature, and the position of the control valve. No lag is assumed except a favorable demand-side capacity lag. Observe that the controller cycles continually, which should be remembered as an inherent characteristic of an off-and-on, or two-position, control.

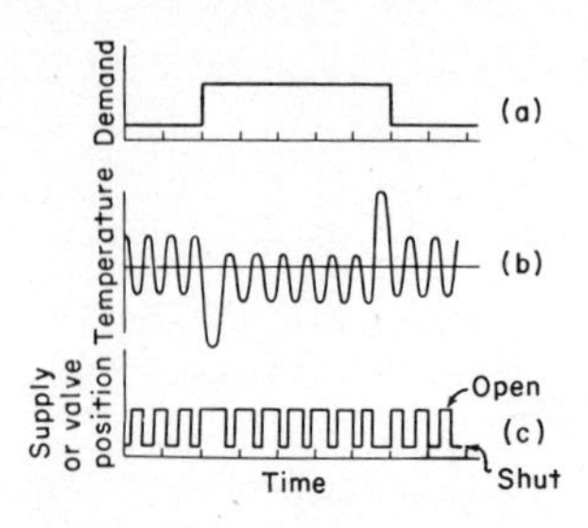

Fig. 8-1 Open-and-shut control characteristics.

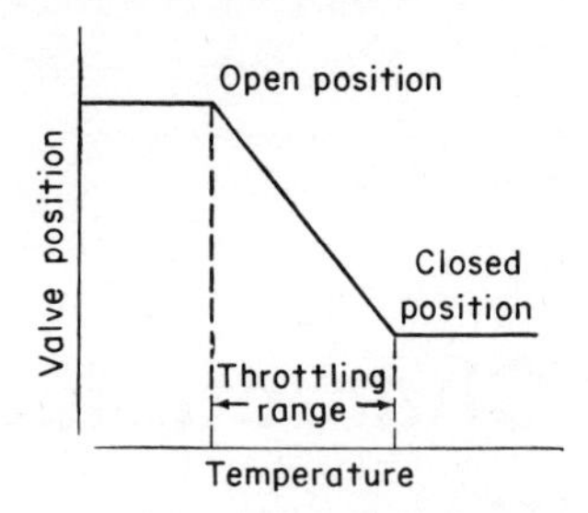

Fig. 8-2 Valve position and temperature relation in proportional-position control.

During the first two units of time the demand is constant, and the temperature of the process cycles evenly about an average temperature line. The amplitude of the cycle defines the control band within which the temperature remains. At times an instantaneous change in the demand takes place, and the temperature drops to a position below the normal position of maximum amplitude and then swings back to a new steady cycle at a lower average temperature. The reason for the large drop at the time of the change in demand is due to the time required to supply the energy lost to the system because of this sudden increase.

After a period of constant demand at the higher level, the demand suddenly returns to its original value, and the characteristic temperature swing reverses itself and returns to a cycle at the same amplitude and position that it had originally. During the period in which the larger demand is being supplied, there is a change in the operating cycle of the valve. At this higher demand the valve is open longer than it is closed. This is because of the change in the storage capacity of the process at the higher rate. The rate of heat transfer also changes at the lower average temperature difference.

The three important characteristics of this type of control are summarized as follows:

1. The control is inherently cyclical in nature.
2. A change in demand produces a wide overswing before even cycling is reestablished.
3. The cycling at the new demand rate is at a different average temperature, lower or higher, depending upon which direction was taken by the demand change.

Proportional Control

A graphic representation of characteristic control obtained with simple proportional control under the same conditions as described previously is shown in Fig. 8-2. Since there is only one position of the control valve for each temperature in the process, only one graph is necessary because it can be made to represent both the controlled temperature and the valve position.

This type of control differs from the off-and-on type in that it may reach a state of temperature equilibrium and continue in this state until a disturbance occurs. In other words, it is not inherently cyclical in character. It can be seen from Fig. 8-3 that the process starts in a position of equilibrium with a constant demand. At time *a* an instantaneous change in demand occurs, and as a result the temperature and valve curves drop from the control point and level off at a new lower temperature after a period of damped, cyclical movement about the new temperature.

The characteristic of the temperature curve made by the process after a sudden change in demand depends upon the width of the control band. Since the valve must function exactly as the temperature that controls it, the deviation from the control point of the new temperature, which produces equilibrium, depends upon the magnitude of the change in demand. The deviation also depends upon the throttling range of the controller because it is larger, with a wider proportional band than with a small band-control range.

The most favorable control is obtained when the throttling range is the smallest that will produce satisfactory control under any particular change in demand. If the throttling range is decreased beyond a certain critical point, the process does not level out into a state of balance but cycles about a point above or below the control point, the exact position of this new average cycling temperature depending upon the throttling-range setting. Control of this type is shown in Fig. 8-3, curve *d*. If the throttling range is increased above this critical point, the temperature curve does not cycle but "wanders" from the control point an unnecessary distance, depending upon how much too wide the throttling band is set.

To obtain the best control with this type of controller, set the throttling range at the point that is critical for the maximum demand change

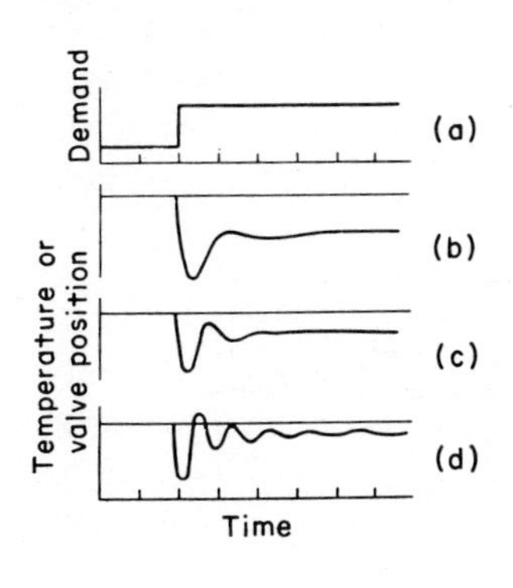

Fig. 8-3 Proportional-position control characteristics.

expected or permitted to prevent cycling and to give the minimum deviation of the process temperature from the control point. In this type of control, as well as in the off-and-on type, any change in demand automatically establishes the process temperature at a new temperature above or below the control point. With off-and-on control the new temperature cycles; with proportional control it reaches equilibrium at this new control point. In either case the new control temperature established may be too far away from the original control point to give the desired results from the standpoint of process chemistry. Where small deviations are not objectionable and where the lags are favorable to control, these modes of control may be satisfactory. Where a very wide throttling band is necessary to eliminate cycling, the control may fail. The characteristics of proportional control may be stated as:

1. Any change in demand requires a change in the temperature necessary to establish equilibrium, hence a new control point.

2. Optimum performance is obtained when the throttling range is set at a critical point below which the temperature cycles and above which it wanders an unnecessary amount.

Floating Control

The curve resulting from a floating control is shown in Fig. 8-4. The three types of floating controls previously described are shown on the same set of coordinates. Since the valve movement is independent of the degree of demand change, it is not necessary to show the demand curve. The direction of the movement is determined only by the direction of the change in demand. Curve *a* represents a simple single-speed floating control; curve *b*, a two-speed floating control; curve *c*, the straight proportional-speed floating control. These curves are self-explanatory and represent a control inherently cyclical in nature, as in the off-and-on type. This type of control can be used only under the most favorable conditions where the speed of the valve is slow and the response of the process rapid.

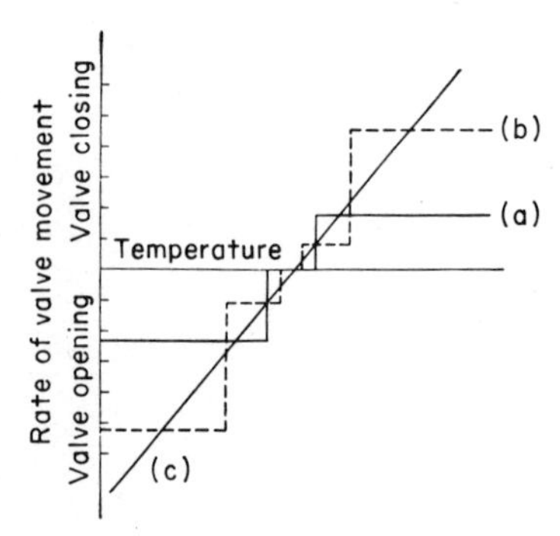

Fig. 8-4 Floating-control rate of valve movement with (a) single-speed control, (b) two-speed control, (c) proportional-speed floating control.

Proportional-and-floating Control

Where it is not permissible for the process temperature to deviate for any appreciable period from the original set point, it is necessary to use a mode of control previously described as proportional and floating. This type of control, seen in Fig. 8-5, requires five curves to explain the characteristics of the system: demand curve *a*, resulting temperature *b*, the movement of the valve due to proportional-position control *c*, the movement of the valve due to floating control *d*, and the resultant movement of the valve under the action of these two forces *e*.

Curve *b* shows the temperature of the medium under control, and it can be seen that the characteristic of the deviation and return to balance are similar to that obtained with proportional-position control. There is a temporary deviation from the control point, a damped cyclical wave, and a final state of equilibrium under the new demand condition. The one significant exception, and the reason for this mode of control, is that the balanced state is reached on the control point and not above or below it. This is close to perfect control insofar as it can be obtained with an automatic device. Figure 8-6 is a graphic illustration of another proportional-and-floating control, showing the normal temperature for a constant demand remaining exactly on the control point which is in the center of the control (or throttling) band. At (*a*) a new demand is suddenly made, and the temperature temporarily deviates from the control point, but owing to the corrective action of the two forces previously mentioned, the temperature promptly returns to the control point. In this case adjustment has been made to produce

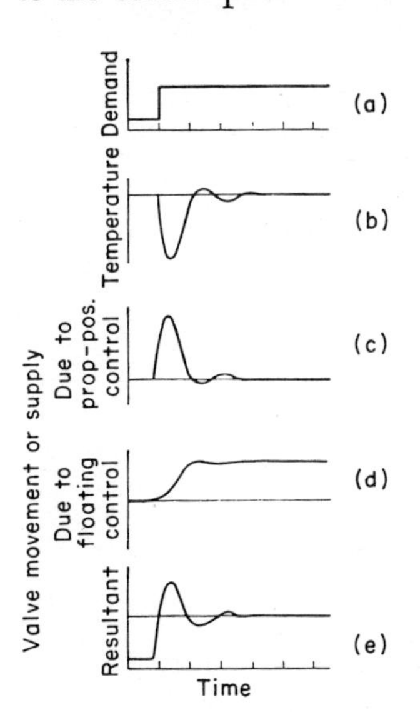

Fig. 8-5 Graphic analysis of floating and proportional control.

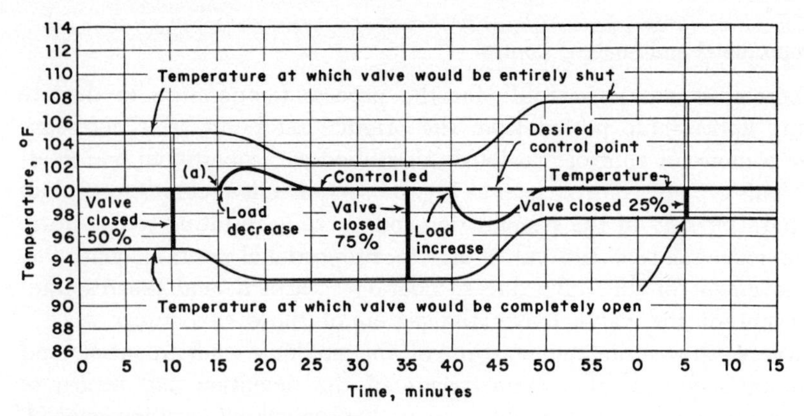

Fig. 8-6 Character of reset control.

asymptotic return rather than damped cyclical return to equilibrium. Note that the control point is no longer in the center of the throttling range but has moved to a new position above the center. This means that the valve is completely open sooner if the temperature rises and is partially open over a greater temperature range below the control point. This action is logical since with the greater demand a larger amount of energy must be supplied each second to hold the temperature constant, hence the nearer-open position of the valve at the point of temperature control.

The adjustment of the floating-and-proportional control is achieved by the use of two secondary adjustments, one controlling the throttling range of the proportional-control force and the other the speed of the floating control.

Figure 8-7 represents graphically the effects of varying each of these two adjustments. In group A the effect of a reduction in the width

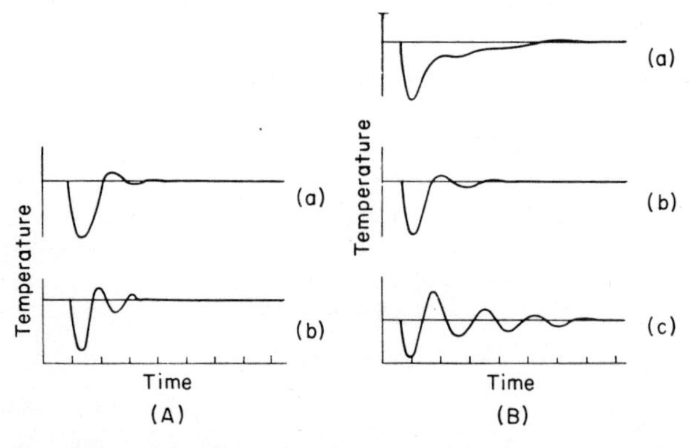

Fig. 8-7 (A) Effect of reducing width of control band. (B) Effect of increasing the floating speed.

of the throttling band is shown. This factor produces a response similar to that produced with simple proportional-position control, except that the cycles move about the original control point. The same critical point is reached, however, where a continuous cycle results if the band is too narrow. Group B shows the effect of an adjustment made on the speed of the floating control. Here too slow a speed retards the return of the temperature to the control point, and too fast a speed produces a cycle of large amplitude about the control point. This cycle can be damped as shown in curve *b*, or it can, if the speed is too fast, produce a permanent cycling temperature.

Adjustment of a floating-and-proportional, or "automatic-reset" (as it is often called in the industry), controller resolves itself into the manipulation of these two adjustments until a minimum temperature deviation is obtained. Figure 8-8 shows a temperature chart of a

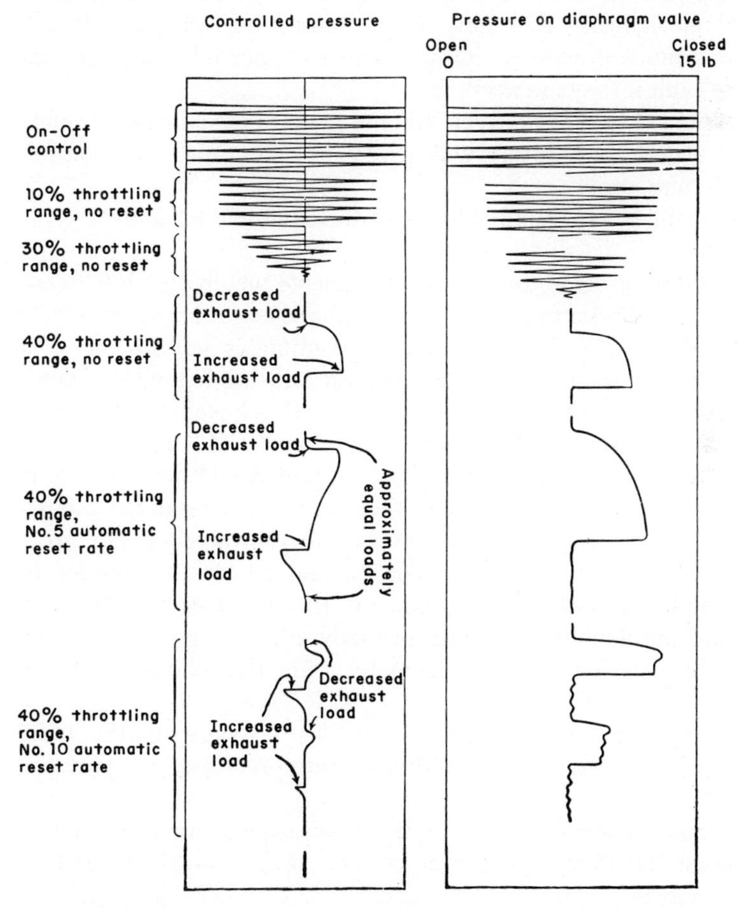

Fig. 8-8 Effect of the type of control and its adjustment when controlling exhaust steam pressure.

recorder giving common disturbances in the control and showing the steps taken to correct them. The cardinal rule for adjusting an automatic reset controller is to move slowly. No further adjustment should be made until the process has had time to reach a state of equilibrium after any previous adjustment. The problem of adjustment will be more completely covered in a subsequent section.

General Discussion of the Problem of Control

There has been much discussion of the relative merits of automatic and manual control and the mode of control that results from normal and automatic adjustments of the controls. A discussion of this kind can become complicated because there are innumerable possible modes of control that might be achieved by normal operation or by automatic-control instruments if there is a good reason for such special control. The control mechanisms that have been studied are adequate for all chemical continuous processes. Other more complicated modes of control might be built if they are needed.

Where a choice is made between automatic and manual control solely on the basis of the method that will achieve the more accurate control, the following rules apply:

1. If the changes are rapid and repetitive, automatic control gives the better results.

2. Where there is time to make adjustments manually, intelligent human control is superior to any form of automatic control.

The first rule is obvious: if rapid and continuous adjustment of a controller must be made, the human operator will lag behind the automatic in making a correction. Furthermore, if adjustments must be continuous, the operator soon tires and loses control. Where any repetitive move must be made continuously over a period of time, a machine does a better job. Conversely, where there is time to make the adjustments necessary for control and where the labor involved is not great enough to tire the operator, intelligent manual control is superior to the most complex automatic controller. It is sometimes thought that a proportional-and-floating controller properly adjusted gives more perfect control than can be obtained manually. For the following reasons this is not true:

1. An automatic controller cannot adjust its response on the basis of experience obtained from intelligent interpretation of previous deviations.

2. An automatic controller must make an adjustment for every deviation. It cannot "wait" to "see" the magnitude of the disturbance before making a correction.

The first of these inherent difficulties with automatic control can be illustrated by a description of the mode of speed control that can be obtained in an automobile with automatic throttle control and with intelligent human control. With automatic throttle control operated by deviations in speed from a set point, the speed of the automobile must deviate from the controlled speed before any corrective action can take place. If the hills that cause the deviation are all the same grade, the controller can be set to return the machine to the correct speed in the minimum possible time. However, if the hills vary widely in grade, it is necessary to set the controller with a throttling range and floating-control adjustment that prevents permanent cycling when the car is on the steepest grade. With this setting the return of the control point of the machine is not so rapid on the small grade. In other words, where the automobile (or any other machine or process to be controlled) is subject to disturbances that vary widely in magnitude and time, the control may be unsatisfactory.

Conversely, the driver of the automobile may have driven the road often enough to know the size of the grade on each hill to be climbed and with the knowledge can open the throttle just enough to prevent the car from losing speed and to keep the speed nearly constant. This mode of control amounts to the automatic control studied plus an optimum adjustment of the throttling range and floating control made for each deviation. Control of this kind would be exceedingly difficult, if not impossible, to achieve automatically.

The second reason for the inability of automatic control to duplicate intelligent manual control on a process in which the deviations vary widely in both magnitude and time is the necessity for the automatic controller to make a response for each deviation. It may be more desirable to wait to see if the deviation is going to be of any magnitude before any change in the control is made. An intelligent operator with knowledge of previous disturbances can learn to tell when a correction should be made. As a result the hand control shows closer regulation than automatic control in many cases.

However, the problem of manual control is complicated by the human equation which exists in all manual operations. It is one thing for an operator to be capable (by constant attention to the controller) of improving slightly on the performance of an automatic controller and another thing actually to produce this superior control. The operator may become careless, leave his station temporarily, or make an error in judgment. The automatic controller is always on the job and responds in a logical way systematically to all disturbances. These advantages of automatic control usually far outweigh the superior control obtained by intelligent manual operation.

Mathematical Analysis of Process Characteristics

Mathematical study of the problem of control is discussed in three parts as follows:

1. The fundamental differential equations of a continuous process automatically controlled
   a. A single-capacity system
   b. A multiple-capacity system
2. An analysis of automatically controlled processes as vibrating systems
3. Harmonic disturbances in an automatically controlled process

It was previously mentioned that an automatically controlled, continuous process does not lend itself easily to a mathematical treatment by means of which quantitative results can be obtained to use in the design of control equipment. The reason for this should be clear from reading the previous sections covering the number of lags that may exist in a system and the characteristics of these lags. Not only are there a large number of variables that can affect an actual process, but these variables are not simple linear functions. Neither is it easy to obtain the heat-transfer coefficients, etc., needed with sufficient accuracy to make a purely mathematical solution entirely reliable.

The characteristics of any automatically controlled process can be expressed with differential equations of varying order, depending upon the number of variables influencing the control. Simple variables influencing the process usually raise the order of the differential equation 1 degree; inertia forces increase it 2 degrees since they are second derivatives with respect to time. The general equation of a controlled system can be an equation of a vibrating system including differentials of a higher order.

Since only linear differential equations are capable of being readily solved, it can be seen that a continuous process can be, and usually is, too complex to lend itself to being solved directly in terms of its variables. This complexity makes it necessary to approach the problem by means of a hypothetically controlled system that resembles an actual practical process but has been simplified by ruling out as variables some of the factors that make the analysis too complicated. The resulting solution of the problem always shows something of the characteristic of the control and may in some causes be almost identical. It is always necessary to remember, however, that the results of this simplified analysis are rigorously true only under the assumptions made. For example, if the solution presupposes that there is no transportation lag, no inertia in the controls, and no transfer lag, it must be considered only as a guide to the possible characteristics of the control since in actual practice, one or more of these lags are sure to be present.

Despite these difficulties, the mathematical approach to the problem will aid greatly in obtaining fundamental knowledge of control characteristics; and as the science of control develops, this phase of the problem will no doubt be more thoroughly mastered. A certain similarity exists between a controlled continuous process and a complex electrical circuit. It may be that the mathematical technique of solving control problems will eventually be as adequate as the means now available in the electrical field.

Fundamental Differential Equations of an Automatically Controlled, Continuous Process

An understanding of the control problem can be gained only from a consideration of the problem as a complex integrated unit. The emphasis up to this point has been on separate units of the process. The characteristics of the controller itself has been covered; the thermometric lag has been studied; the damping characteristics of the measuring instrument have been considered; and mention has been made of the characteristics of fluid lags in the signal-transferring system. All these studies are important as an aid to a complete understanding of automatic control and as a help in the proper design of the parts that make up the complete system.

The analysis that follows will cover the simple-unit, continuous process and will follow the mathematical analysis of C. E. Mason,* using the nomenclature of the original study made by Spitglass,† Mitereff,‡ Stein,§ and others. All these studies approach the problem first as a simple continuous process with only capacity lag and then build up the more complex equations resulting from the addition of other capacities in the system.

Analysis of a System with Capacity Lag

Figure 8-9 represents a simple-unit continuous process having only capacity lag. It is assumed that the heat supplied is an electrical coil without lag and that there is sufficient agitation to maintain a constant

* C. E. Mason, "Quantitative Analysis of Process Lags," *Trans. ASME*, Pro-60-1, 1938, pp. 327–334.

† Albert F. Spitglass, "Quantitative Analysis of Single-Capacity Processes," *Trans. ASME*, Pro-60-9, November 1939, pp. 7–49.

‡ Sergei D. Mitereff, "Principles Underlying the Rational Solution of Automatic Control Problems," *Trans. ASME*, FSP-57-9, 1935, pp. 159–163.

§ T. Stein, "Selbsreglung, ein neuses Gesetz der Regelteehnik," *Ver. Deut. Ingr. Z.*, vol. 72, Feb. 11, 1928, pp. 165–171.

temperature throughout the vessel. It is further assumed that there is no heat loss by radiation. The nomenclature used in the analysis is as follows:

A_a = heat capacity of the vessel
T_a = temperature existing in the vessel under normal balance
Q_a = heat flow in Btu per min leaving vessel
T_P = potential temperature that would be reached under any set of conditions if no change of any kind were made
Q_0 = amount of heat entering the vessel in Btu per min
Q_s = supply of heat added to control the temperature
T_0 = temperature of the vessel when $Q_s = 0$
R_a = resistance factor defined as T_a/Q_a
t = time measured from time disturbance occurred, in min
T'_a, T''_a = first and second derivatives with respect to time
T_a0, T_a0' = initial values of these factors when $t = 0$

It is first assumed that the process is in a state of balance with the material flowing into the vessel at a temperature T_0 and with a heat content equal to Q_0. If it is further assumed that the temperature T_0 is 0°F and Q_0 is the quantity of heat existing above a temperature of 0°F, the only heat leaving the vessel above the basic zero temperature will be that added by the controlling heat source.

Then
$$Q_s = Q_a \tag{8-1}$$

and the temperature will be constant.

An instantaneous change in the demand will cause a change in the amount of heat flowing into the vessel; and since Q_s is constant, the temperature in the vessel will start to rise or fall, depending upon whether more or less heat is flowing to the vessel. This change in temperature is at a rate expressed by

$$T'_a = \frac{Q_s - Q_a}{A_a} \tag{8-2}$$

If we hold Q_s constant in the face of this rise in temperature and if the amount of heat flowing out of the vessel is not affected by the

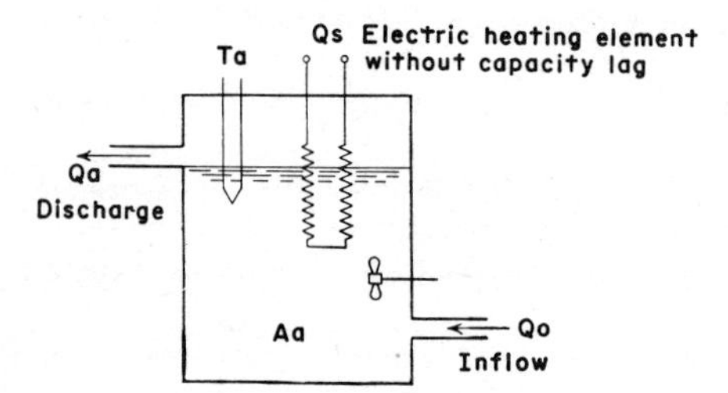

Fig. 8-9 Simple continuous process with only capacity lag.

change in temperature, the temperature at any time can be expressed as

$$T_a = T_{a0}'t + T_{a0} \tag{8-3}$$

Actually the condition in this equation would not exist since the change in the amount of heat entering has an effect on the amount leaving the vessel. If this were not so, the temperature in the vessel would increase indefinitely. This temperature rise or fall starts at a high rate because of the large temperature differential and gradually slows down as critical temperature T_P is reached where equilibrium in the system is again established between the heat entering and leaving the vessel. This potential temperature is described by A. Ivanoff* and is an important concept in automatic control because it determines the inherent controllability of the process.

The inherent ability of a system to control itself (its self-regulation) is of fundamental significance in the problem of control. A simple explanation of this property may be obtained from Fig. 8-10 which shows a hydraulic system with a liquid flowing in one side and out the other. If more fluid is pumped into the vessel in a given time, the liquid level will rise, and since there is greater pressure on the outlet orifice, there is an automatic increase in the fluid flowing out of the vessel. A head pressure will eventually be established that is sufficient to discharge as much fluid as is being pumped into the vessel, and an equilibrium is established at the larger demand. This self-regulating tendency may be great enough to satisfy the process without any additional control equipment. However, other requirements may make the self-regulation inoperative. For example, the vessel in Fig. 8-10 may overflow before sufficient head is established or in the case of a furnace, the walls of the furnace may burn up before equilibrium is established. However,

* A. Ivanoff, "Theoretical Foundations of the Automatic Control of Temperature," *J. Inst. Fuel*, vol. 7, no. 33, February 1934, p. 117.

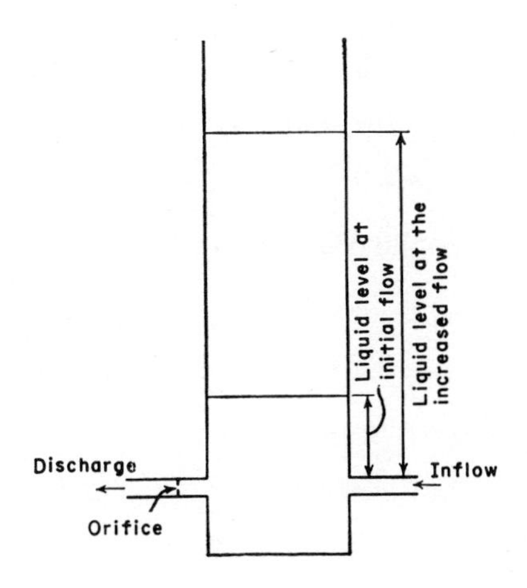

Fig. 8-10 Self-regulating characteristic of vessel under hydrostatic pressure.

a potential balance exists in any set of conditions, and this potential temperature is of great aid in the mathematical analysis of the control problem.

Quantity Q_a is a variable in the process, and Eq. (8-4) must be modified to take this fact into account if the expression is to represent the true temperature. Ratio T_a/Q_a represents resistance factor R_a which affects the flow of heat from the vessel, or

$$R_a = \frac{T_a}{Q_a} \tag{8-4}$$

and

$$Q_a = \frac{T_a}{R_a} \tag{8-5}$$

then

$$T'_a = \frac{Q_s - T_a/R_a}{A_a} \tag{8-6}$$

or in a different form

$$T'_a = \frac{R_aQ_s - T_a}{R_aA_a} \tag{8-7}$$

since

$$R_a = \frac{T_a}{Q_a} \tag{8-8}$$

Under any balanced condition we can also write

$$R_a = \frac{T_P}{Q_0 + Q_s} \tag{8-9}$$

since at the potential temperature the quantity of heat leaving the vessel is equal to $Q_0 + Q_s$. Then if Q_0 equals zero, the equation can be written

$$T_P = R_aQ_s \tag{8-10}$$

Substituting this in Eq. (8-7), we have

$$A_aR_aT'_a + (T_a - T_P) = 0 \tag{8-11}$$

with T_P, A_a, and R_a as constants. Integrating this equation,

$$T_a - T_P = C_ae^{-(1/A_aR_a)t} \tag{8-12}$$

The constant of integration C may be determined by setting $t = 0$.

Then

$$(T_a - T_P)_0 = C_ae^{-0} \tag{8-13}$$

or

$$C_a = (T_a - T_P)_0 \tag{8-14}$$

then

$$T_a = T_P = (T_a - T_P)_0e^{-(1/A_aR_a)t} \tag{8-15}$$

or

$$T_a = T_P + (T_a - T_P)_0e^{-(1/A_aR_a)t} \tag{8-16}$$

This is the fundamental equation for the temperature in the vessel at any instant after a disturbance in the flow of fluid has taken place. The equation is logarithmic in form and expresses the characteristic lag of a thermometer bulb suddenly immersed in a bath at a higher temperature. Here T_a represents the temperature of the thermometer; T_P, the temperature of the bath; and A_aR_a, the coefficient of lag. Therefore factor A_aR_a has the dimension of time just as a thermometer has it.

The temperature at any time can therefore be expressed in terms of the initial temperature, the potential temperature, the capacity of the vessel, and two constants. This equation may be put in its logarithmic form and solved for time as follows:

$$t = R_aA_a\left[\log_e\frac{(T_a - T_P)_0}{(T_a - T_P)}\right] \tag{8-17}$$

If constant A_a and temperature T_a and T_P are known, constant R_a can be determined for the value of t producing potential temperature T_P. Once constant R_a is obtained, the time required for the system to reach any other temperature may be obtained.

This time can then be used as a guideline in selecting the control equipment because the speed and character of the response mechanism can be checked against the allowable time for a response if the process is to be kept within prescribed limits.

Analysis of a System with Capacity and Transfer Lag

Capacity and transfer lags are the two most likely to occur in any temperature-control problem. Unfortunately the addition of another lag makes the analysis much more difficult. However, it is possible to study the system as follows. Figure 8-11 shows a simple process similar to that in Fig. 8-9, except that the heating element is not in intimate contact with the fluid to be heated but is enclosed in a vessel that causes transfer lag. A study of this process will require the addi-

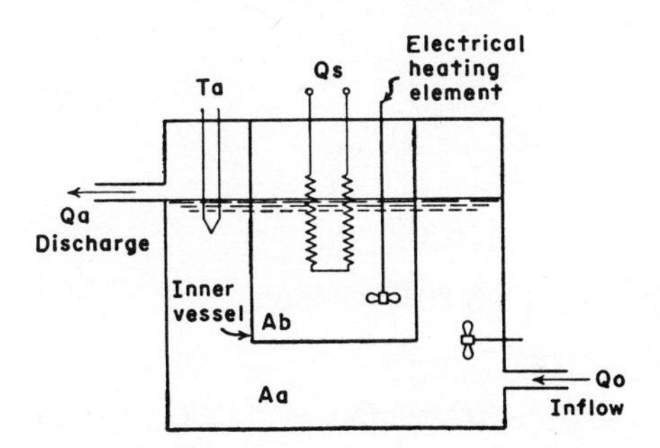

Fig. 8-11 Simple continuous process with capacity lag and transfer lag.

tional items

A_b = heat capacity of the inner vessel
Q_b = rate of flow of the heat from A_b to A_a
T_b = temperature in A_b
R_b = ratio of the temperature drop between A_b and A_a to the flow of heat Q_b at the instant $R_b = T_b - T_a/Q_b$

If a sudden disturbance is made in the demand in this system, the rate of temperature rise in the vessel is

$$T'_a = \frac{Q_b + Q_0 - Q_a}{A_a} \tag{8-18}$$

and the rate of temperature rise in the inner vessel is

$$T'_b = \frac{Q_s - Q_b}{A_b} \tag{8-19}$$

As previously shown

$$T_a = R_a Q_a \tag{8-20}$$

Then

$$T_b = T_a + R_b Q_b \tag{8-21}$$

and when these last two equations are combined,

$$T_b = R_a Q_a + R_b Q_b \tag{8-22}$$

Then by differentiating Eqs. (8-20) and (8-21), we have

$$T'_a = R_a Q'_a \tag{8-23}$$

$$T'_b = R_a Q'_a + R_b Q'_b \tag{8-24}$$

Equating the values of T'_a in Eqs. (8-18) and (8-23), we have

$$A_a R_a Q'_a = Q_b + Q_0 - Q_a \tag{8-25}$$

Equating the values of T'_b as given in Eqs. (8-19) and (8-24), we have

$$A_b R_a Q'_a + A_b R_b Q'_b = Q_s - Q_b \tag{8-26}$$

Solving Eq. (8-25) for Q_b, we have

$$Q_b = (Q_a - Q_0) + A_a R_a Q'_a \tag{8-27}$$

Differentiating this, we have

$$Q'_b = (Q_a - Q_0)' + A_a R_a Q''_a \tag{8-28}$$

Substituting Eqs. (8-27) and (8-28) in Eq. (8-26), we have

$$A_b R_a Q'_a + A_b R_b (Q_a - Q_0)' + A_a R_a A_b R_b Q''_a = Q_s - (Q_a - Q_0) - A_a R_a Q'_a \tag{8-29}$$

This may then be arranged as

$$A_aR_aA_bQ_b'' + (A_aR_a + A_bR_a + A_bR_b)Q_a' + Q_a - Q_0 = Q_s + A_bR_bQ_0' \quad (8\text{-}30)$$

If flows Q_0 and Q_s are constant during a period of time, $Q_0' = 0$, and the preceding equation reduces to

$$A_aR_aA_bR_bQ_a'' + (A_aR_a + A_bR_a + A_bR_b)Q_a' + (Q_a - Q_0 - Q_s) = 0 \quad (8\text{-}31)$$

Multiplying this equation by R_a, we have

$$A_aR_aA_bR_bT_a'' + (A_aR_a + A_bR_a + A_bR_b)T_a' + T_a - R_a(Q_0 + Q_s) = 0 \quad (8\text{-}32)$$

Then since $T_P = R_a(Q_0 + Q_s)$ and since T_P is a constant, we have

$$A_aR_aA_b(T_a - T_P)'' + (A_aR_a + A_bR_a + A_bR_b)(T_a - T_P)' + (T_a - T_P) = 0 \quad (8\text{-}33)$$

The solution of this differential equation is

$$T_a - T_P = C_be^{k_bt} + C_ae^{k_at} \quad (8\text{-}34)$$

In this equation factors K_b and K_a are the two roots of auxiliary quadratic Eq. (8-33) and not simple values, as were obtained in the solution of the simple single-capacity system. The constants also are complicated by the fact that they involve differentials of T_{a0}. As a result of these difficulties a quantitative solution is difficult though possible. This is contained in Mason's study previously discussed.

Analysis of Automatically Controlled Processes Considered as Vibrating Systems

A second method of studying the problem of automatic-process control has been evolved by Ivanoff* and is now considered, using the nomenclature and mathematical approach presented in the original paper.

This method of studying a controlled system has the advantage of more clearly integrating the entire controlled process into a single equation representing the temperature existing in the controlled vessel at any time. Although the quantitative solution of the problem in this approach is no easier than with the method previously described, the novel method of analysis should help the reader to grasp the fundamentals of the problem.

When the problem of automatic temperature control as an oscillating system of a sinusoidal nature is studied, it is necessary to limit the study to a system with simple capacity lag if the equations are to be

* Ivanoff, *op. cit.*, p. 117.

simple in form. Space does not permit the detailed study of this analysis, but the technique can be shown by considering only the on-and-off type of control in a system with capacity lag only.

Concept of a Sine Wave of Potential Temperature

In this method of analyzing a controlled process it is assumed that the potential temperature, which was studied in the previous analysis by Mason, can be assumed to be sinusoidal in nature under the influence of any deviation from a state of perfect equilibrium in the system. We have seen in the sections on the graphic analysis of process characteristics under the influence of various control mechanisms that in every case this cyclical tendency was exhibited by the controlled temperature, which would indicate the validity of this assumption.

This sine wave of potential temperature is assumed in its characteristics under the influence of the disturbances in the system to vary in three ways: (1) the form of the wave may change, (2) its amplitude may change, and (3) its wave may be displaced along its axis.

If it is assumed that the disturbances in the process do not change the characteristics of the process, the potential temperature can be represented by

$$\lambda = A \sin (mt - u)$$

where λ = potential temperature
t = time
m, A, and u = constants

The actual graph of the temperature in the vessel resulting from this potential wave is distorted by the lags of the process and can be represented as

$$\Theta = A_e^{-\phi(m)} \sin [mt - u - \chi(m)] \tag{8-35}$$

where e = base of natural logarithms
$\theta(m)$ and $\chi(m)$ = empirical functions

Study of the control of a simple automatically controlled process would indicate that empirical functions $\phi(m)$ and $\chi(m)$ take the form $c\sqrt{m}$. This being the case, we have

$$\theta(m) = \chi(m) = c\sqrt{m} \tag{8-36}$$

and Eq. (8-35) can be rewritten

$$\Theta = A_e^{-c\sqrt{m} \sin (mt - u - c\sqrt{m})} \tag{8-37}$$

Equation of Automatic Regulation

Any continuous process that is unstable, owing to changes in either the demand or the supply sources, will vary from the point of original temperature balance if no corrective action is taken. This variable temperature is a function of time and can be written

$$\Theta = F(t) \tag{8-38}$$

If the process is placed under automatic control, the potential temperature is varied by the control mechanism, and the system will actually exhibit a temperature curve that has been corrected by the control. This actual temperature is expressed as another function of time:

$$\Theta = f(t) \tag{8-39}$$

The influence of the controller is then represented at any time as the difference between these two functions. If we call this difference Θ_z, we have

$$f(t) = F(t) - \Theta_z \tag{8-40}$$

Θ_z in this equation is the correction applied by the controller in varying the potential temperature. This factor is dependent upon the characteristics of the process under control and is related to the empirical functions in Eq. (8-35). We can therefore write

$$\Theta_z = \Phi_l[\theta(m),\chi(m),f(t)] \tag{8-41}$$

or if the empirical functions are considered to be included in $f(t)$, the fundamental equation of control can be written

$$f(t) + \Phi[f(t)] - F(t) = 0 \tag{8-42}$$

The form of function $[f(t)]$ depends upon the number and length of the lags in the process and the type of control employed. If the process is complex, this will be a complex function.

Analysis of a Process Having On-and-off Control

For simple on-and-off control the potential temperature has only two values, one representing the on position of the valve and one the off position. If this potential wave is assumed to be symmetrical about the time axis, it can be represented by the harmonic series

$$\lambda = \frac{4}{\pi}\lambda_o\left[\sin(mt) + \tfrac{1}{3}\sin 3(mt) + \tfrac{1}{5}\sin 5(mt) + \cdots\right] \tag{8-43}$$

where $m/2\pi$ = frequency
t = time
λ_o = amplitude of the wave

The temperature that would be recorded because of this potential wave can be obtained by substitution of value Θ given in Eq. (8-35). This is

$$\Theta = \frac{4}{\pi} \lambda_o [e^{-c\sqrt{m}} \sin (mt - c\sqrt{m}) + \tfrac{1}{3} e^{-c\sqrt{m}} \sin (3mt - c\sqrt{3m}) + \cdots] \quad (8\text{-}44)$$

where c = constant representing the coefficient of the particular process. The oscillation of these two curves of actual temperature, shown in Fig. 8-12, is seen to be identical in form with the curves of the controlled temperature and valve position previously shown in the section covering the characteristic control obtained with on-and-off control.

Harmonic Disturbances in an Automatically Controlled Process

When a process involves the combination of several typical controlled units of the type shown in Fig. 8-1, the control problem may be complicated by the introduction of harmonic disturbances in the system. For example, it may be found that to obtain closer control over a process of the type just studied, it is necessary to reduce the fluctuation in the heat supply being used to produce control in the process. If the process contains unfavorable lags within the circuit of the control as shown in Fig. 8-1, it is natural that a secondary fluctuation in the heat supply further increases the difficulty of control. To reduce the amount of this secondary deviation, a control mechanism is often installed on the heat-supply source.

The system may now be considered as two separate but interrelated controlled processes. The fundamental temperature curve exhibits characteristics that depend upon the amplitude and phase relations of the

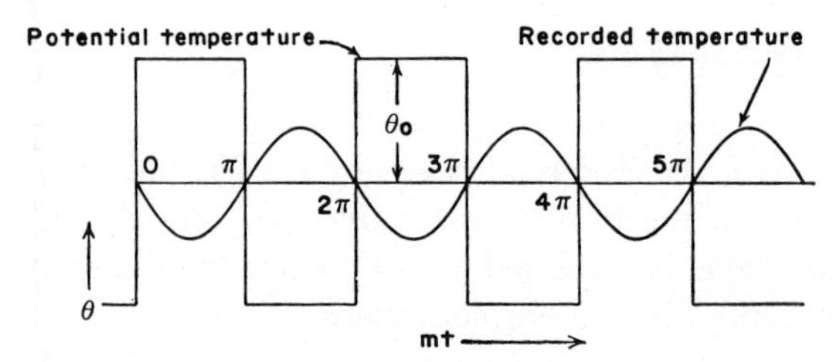

Fig. 8-12 Curves of actual temperature and potential temperature using off-and-on control.

harmonic curves. If these harmonic components are of a form and phase to cancel one another, a stable control results. If, however, they reinforce one another, the control suffers, and violent cycling may result.

This phenomenon of harmonic disturbance is the cause of some mystifying control characteristics in practice. For example, a process will control perfectly for a certain period of time and then suddenly become very unmanageable without reason. Again, the process will be less stable with the secondary control in operation than when this part of the process is left to vary at will or is permitted to cycle widely.

The first characteristic is due to the fact that the process is originally adjusted with the time periods of the harmonic components set to dampen any cycling under the conditions prevailing while the adjustments were made. A change in demand will, unfortunately, change the time period of the cycle of the harmonic wave representing the primary controlled process, as was shown clearly in Fig. 8-1, without affecting the harmonic curve of the heat supply. This may throw the harmonic waves in phase and in this way produce wide deviations in the controlled temperature.

In the second case the control may be closer if the secondary supply source is allowed to fluctuate because it may have a cycle 180° out of phase with the harmonic wave of the primary control. The control of the secondary demand may in this case reduce the magnitude of its deviations but also throw it in phase with the primary harmonic wave. This action is somewhat unlikely but possible if the coefficients of the wave happen to be favorable.

Where these difficulties occur, the only solution is to provide controllers capable of adjustments that will produce harmonic waves either out of phase with each other or of such small amplitude that they have a negligible effect on the control. Here the automatic-reset, or proportional-and-floating, type of control is of great advantage since it is capable of adjustments that will produce harmonic waves of wide variation in time period. This will be illustrated in the case of the liquid-level control covered in Chapter 10. The liquid-level-control system has a primary control mechanism for controlling level and a secondary control for controlling sudden changes in pressure in the vessel that adversely affect the liquid level. Automatic-reset controllers are used for both control mechanisms to avoid harmonic disturbance.

CHAPTER NINE

Automatic-control-valve Characteristics

THE CONTROL VALVES for the control mechanisms studied in the previous chapter have been of two types, the open-and-shut valve and the throttling type. The open-and-shut valve has been described as the control valve used with the simple off-and-on mode of control, and the throttling type is used with the proportional control and the proportional-and-floating control.

Nothing has been said of the characteristics of these valves, although in the case of proportional controllers the assumption was that there is a linear relationship between the percentage of the total valve movement and that of the total flow. This was seen in Fig. 8-3 where the same curve represents both the valve position and the temperature of the controlled medium.

The characteristic of the control valve in any automatically controlled process is of great significance where throttling action is required. This type of valve will be analyzed in detail. The open-and-shut type of valve may have any design or any type of flow characteristics since it moves from a wide-open position to completely closed in too short a time to permit its throttling characteristics to influence the control. As a result a gate, a globe, a needle, or a diaphragm valve, or any other type of valve that is convenient and economical, may be chosen for off-and-on control.

Throttling-valve Characteristics

The control mechanisms used for automatic control of the proportional type are designed to produce a linear relationship between the change in controlled temperature and the force applied to the valve to affect control. They are also designed to produce a linear relationship between the rate of change of the process variable and the rate of change of the valve-positioning force. If these two relationships exist, good control results, provided there is also a linear relationship between the percentage of total valve movement and the percentage of total flow resulting from any change in valve position.

In a system requiring throttling control it can be said that of all the characteristics contributing to the stability of the control, the most important is the flow characteristics of the valve control. If this valve does not produce a flow response of the desired magnitude required by the controlled mechanism, accurate temperature or flow control is impossible. Therefore it is necessary to design a control valve to match the characteristics of the control mechanism.

Rangeability of a Control Valve

It can be seen in Fig. 9-1 that curves 1, 2, and 3 cannot control flow throughout their entire range because when X is equal to zero, the valve is still passing a small percentage of the fluid. In other words, the valve does not close completely at the bottom position of the valve stem. The ratio of the maximum to the minimum controllable flow is known as the "rangeability of the valve." For example if X is equal to 5 percent when Y is zero, the rangeability factor of the valve is

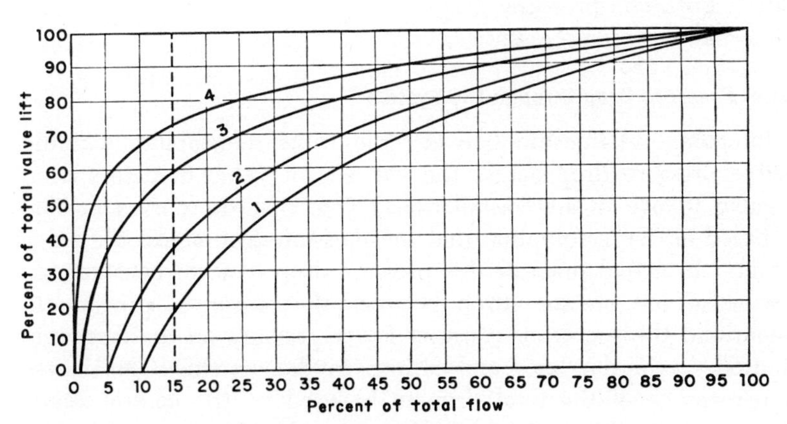

Fig. 9-1 Flow characteristics of valve with log curve. *(The Foxboro Company)*

20. The valve must always have sufficient rangeability to dominate the control; it must be able to close far enough to counteract any over-range in the process and to open far enough to supply any deficiency.

From the fundamental expression for flow through an orifice, it is known that the valve area is directly proportional to the flow and inversely proportional to the square root of the pressure drop across the valve. From this fact the allowable pressure drop across a valve can be ascertained if the rangeability factor and the minimum permissible rangeability of the valve are known. Since the rangeability factor represents a valve area, we may write

$$\frac{\text{Rangeability of the valve}}{\text{Rangeability required by the process}} = \sqrt{\text{pressure-drop rangeability}}$$

If, for example, the rangeability of a valve is 16 and minimum rangeability required by the process is 4, then

$$(16/4)^2 = 16$$

which is the rangeability of the pressure drop that may be permitted and still maintain the valve in control.

The rangeability of a valve can be combined as follows with the rangeability of pressure drop:

$$P_r = \frac{V_r^2}{F_r}$$

where P_r = max P_1—min P_2/min P_1—max P_2
F_r = max flow/min flow
V_r = rangeability of control valve
P_1 = upstream pressure
P_2 = downstream pressure

Effect of Pressure Drop through the System

The foregoing relationship between the valve rangeability and the permissible pressure drop across the valve is of value in determining which valve to use in any installation. Note that the curves in Fig. 9-1 are based on the assumption that the pressure drop across the valve is constant. In actual practice this pressure drop is never constant because some of the pressure drop between the source of pressure in the system and the source of constant lower pressure on the discharge side is used up in the lines and fittings in the system. This leaves only a percentage of the total drop to be used by the control valve. Also, the remaining drop used up in the system by the lines and fittings varies as the square of velocity of flow according to the hydraulic theory and therefore is different for each setting of the valve. This fact tends to upset the desirable flow characteristics of the valve, as seen in Fig. 9-2, which represents the actual characteristics of a valve in a system with a variable-pressure drop. This is a valve with a parabolic flow curve and with the total pressure drop available at the valve presenting good characteristics. However, as E (the percentage of the total drop effective at the valve) is decreased, the characteristics become less desirable since equal percentages of valve movement no longer produce equal percentages of increase in flow.

The obvious course in the design of automatic-controlled process equipment is to arrange as small a pressure drop through the lines and fittings as possible. Thus the desirable flow characteristics of the valve are preserved. A good rule in the design of lines that have control valves in them is to keep the liquid velocity below 5 ft per sec. The problem of pressure drop is of great significance where the control is obtained with very small pressure drop. In this case any pressure drop in the lines is serious because it represents so large a percentage of the total available drop. For example, if the control has a normal pressure drop of only 2 lb, an additional variable-line-pressure drop of only 1 lb will seriously affect the control if it does not nullify it entirely.

Where the problem of line-pressure drop cannot be solved in any other way, it is possible to install a control valve just ahead of the valve operated by the control mechanism and designed to maintain a constant-pressure drop across the valve. This valve is operated by the difference in pressure on the upstream and downstream sides of the control valve through pressure taps taken from either side of the control valve and operating the diaphragm of the constant-pressure valve. An arrangement of this kind is shown in Fig. 9-3.

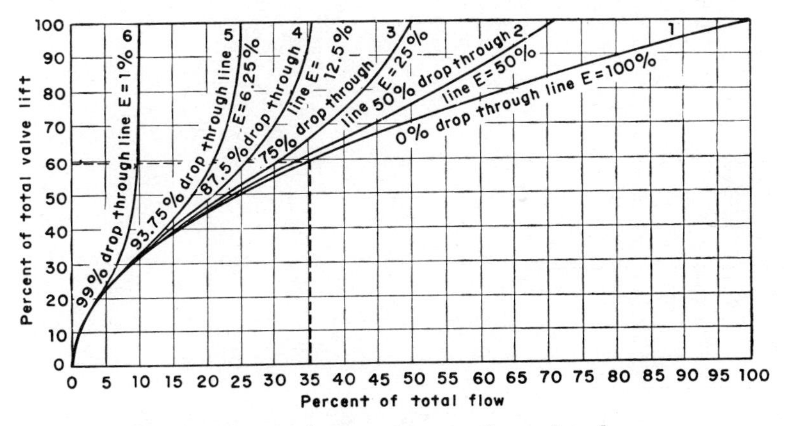

Fig. 9-2 Effect of pressure-drop variation on flow valve characteristics. (*The Foxboro Company*)

Valve-port Design

The theory with which we have been dealing has been based on the fact that the valve characteristics conform to the equation

$$Y = C \log AX$$

This is the correct curve for a valve with correct flow characteristics throughout its entire range, and it has been shown that the constants of the equation could be arranged so that they give a rangeability of more than 100 percent of the valve-stem travel if desired. In practice it is impractical for the following reasons to construct a valve with 100 percent rangeability:

1. A minumum clearance is needed between the valve seat and plug to avoid binding and to allow for expansion and contraction.
2. If steam is used in the valve, wire drawing would be excessive near the closed position of the valve.

Usually the valve is designed with a rangeability covering between 95 and 98 percent of the total range of the valve opening, depending upon the size of the valve. Therefore, the constants of the valve are chosen to permit the curve to cross the x axis at a value somewhere between 1 and 8 percent of the total flow.

The design of the flow-control valve is usually similar to that shown in Fig. 9-4. This type of valve is double-ported and semibalanced, either with a v port as shown or with a parabolic type of port shown in Fig. 9-5. The double-ported valve is chosen because it lends itself to control with an air-operated diaphragm (because the pressure of

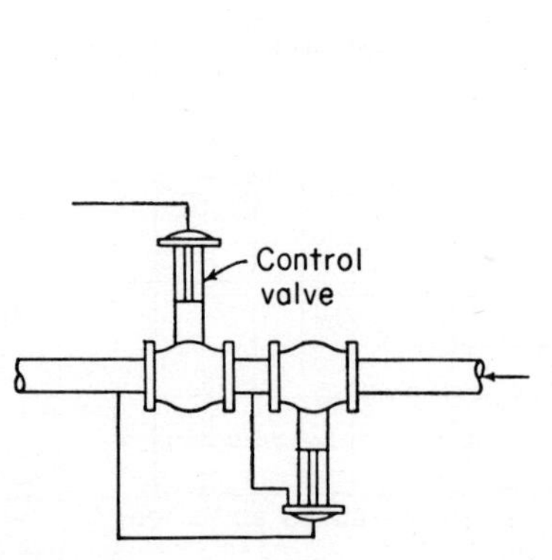

Fig. 9-3 Constant-pressure valve maintaining a constant pressure-drop across control valve.

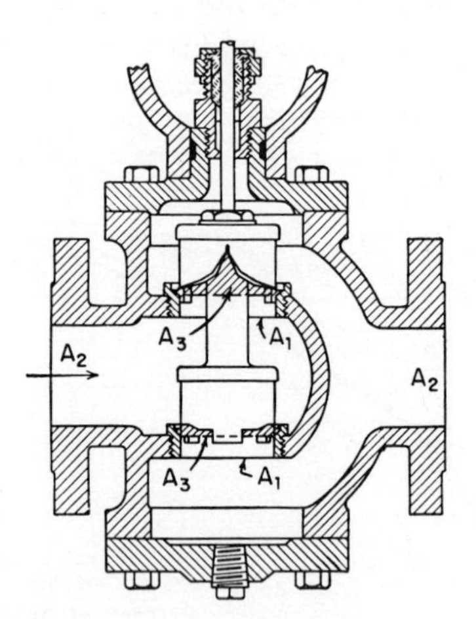

Fig. 9-4 Double-ported valve body with plug at maximum lift. (*Honeywell, Incorporated*)

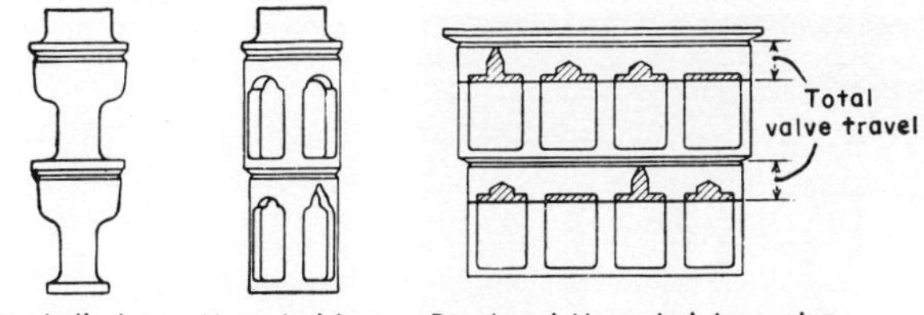

Fig. 9-5 Control-valve plugs.

the plug is balanced except for the small area of the valve stem, and the valve will therefore move freely in either direction with a minimum application of power from the diaphragm).

The ports used are usually of the *v* or the parabolic type, both of which are shown in Fig. 9-5. The simplest to construct is the *v*-port type (triangular), with the characteristic that its area-altitude curve is parabolic in shape.

The parabolic curve of area with respect to height of the triangular opening produces characteristics almost identical with that of a valve having perfect characteristics, provided the origin of the parabola is chosen correctly. The equation of a parabola is

$$Y^2 = AX$$

and if the curve is transposed on the x axis, a proper amount and the proper constant assigned to it, it can be made to cross the curve of the true log curve of flow in three places. In this case the error, or deviation, of the parabolic curve from the true curve of correct flow is very small, as can be seen from the curve of Fig. 9-6.

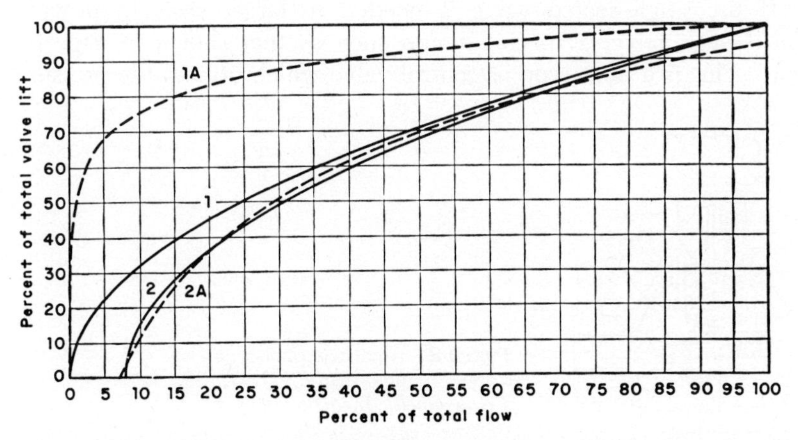

Fig. 9-6 Comparison between parbolic curve and log curve crossing at three points. (*The Foxboro Company*)

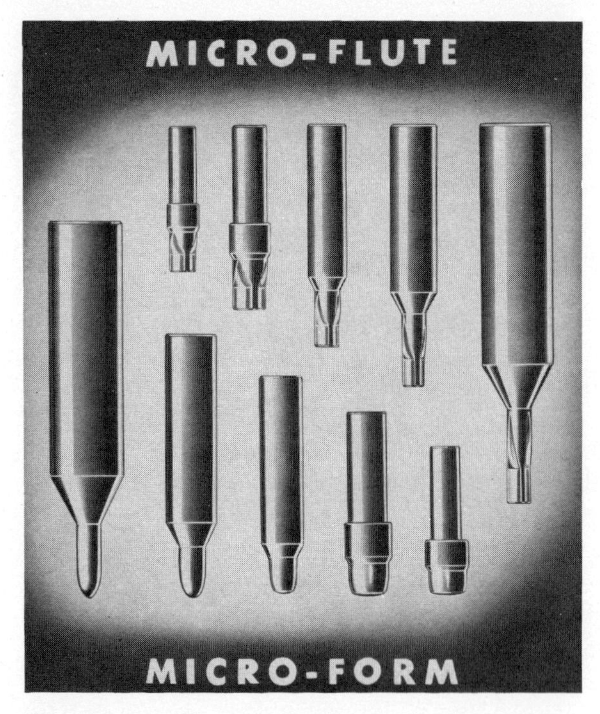

Fig. 9-7 Single-seated valve of various designs. (*Fisher Governor Company*)

Occasionally it is desirable to have a single-seat valve because of the cost or where the control is not critical. Several designs are shown in Fig. 9-7. Such valves are used where the port area is ¾ in. or less. If the single-seated valve is needed in larger sizes to provide a tight-closing valve, a special design such as that shown in Fig. 9-8 is used. This is a pilot type of control valve which utilizes the pressure

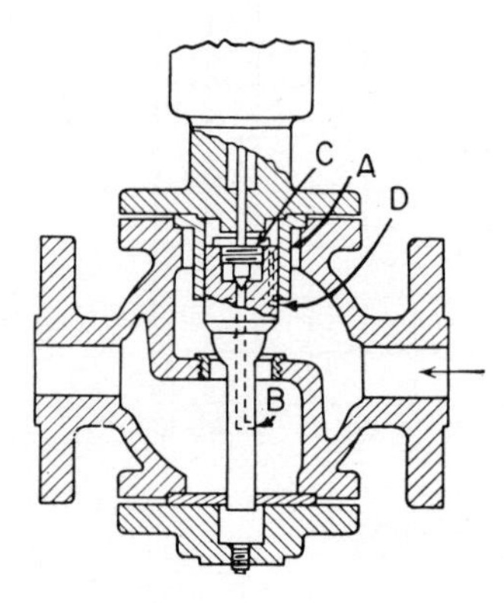

Fig. 9-8 Auxiliary pilot-type valve, single-seated body. (*Honeywell, Incorporated*)

in the valve to provide the large initial force required to move the single-seated type of valve from its closed position. Operation of the valve is by means of a small secondary valve operated by the valve stem. When the valve starts to open, the steam flows into the lower side of chamber C from the low-pressure side through port B and from the high-pressure side to the top of the chamber through port D. Since port B is larger than *d*, a higher pressure is built up on the lower side of the disk separating the two ports, and this force tends to open the valve.

Effect of Bypass Flow on Throttling Control

For the following reasons, where true throttling control is needed in a continuous process, it is never desirable to arrange a bypass about the control valve to carry some of the flow:

1. The total flow resulting from that passing through the control valve and bypass does not conform to the perfect percentage curve desirable for good control.
2. The rangeability of the valve is greatly reduced.

Figure 9-9 represents the effect of bypassing 50 percent of the flow through a bypass valve. The true percentage curve at this point is approaching a straight line, and the actual flow characteristics are similar to those with the origin at or near zero, except that the curve crosses the axis at a value of 50 percent. A wide deviation from the true percentage curve results from the move. A marked reduction in the rangeability of the valve is also produced by use of the bypass valve. The rangeability factor is 50 with the origin of the curve at 2 percent but is only 2 when 50 percent of the flow is bypassed.

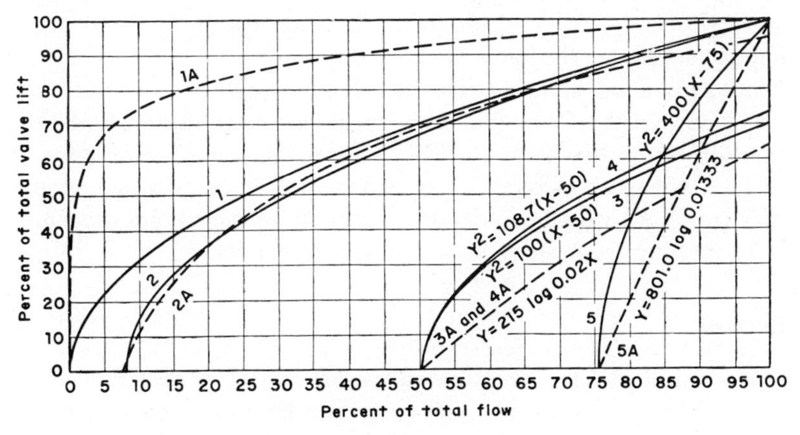

Fig. 9-9 Effect of bypass valve on flow characteristics. (*Honeywell, Incorporated*)

A bypass valve is effective in improving the throttling characteristics of a valve only where the ordinary globe type of valve is used. In this case adjustment of a bypass may improve the control somewhat where the system requires only a narrow band of rangeability. In the case of an open-and-shut type of control a bypass is often helpful in establishing the magnitude of the fluctuations in flow produced by the control. Where true throttling action is needed, the only solution is to use a valve of the correct size without bypass adjustments.

Calculation of Valve Sizes

A general rule to follow in choosing the size of the control valve for a process is to keep the valve no more than half the size of the line. If it must be as much as three-quarters the size of the line to pass the desired volume of fluid, the control will be adversely affected. The optimum rule for this is to compute the valve size on a basis of double the normal flow through the system plus 10 percent. Using this size of valve for control, design the lines in the system twice this size in order to reduce the line-pressure drop to a minimum.

In actual practice manufacturers of control valves provide nomographs, or slide rules, to aid in calculating the correct valve size. In general it will not do to make an arbitrary assumption of the pressure drop that will be available at the point where the valve is to be placed and to proceed from there to determine the valve size. To be sure of the correct size, one must analyze the process completely, computing the pressure losses from point to point until a rational value is arrived at for the pressure drop available at the valve. An additional factor of good judgment is of great help in evaluating the valve size arrived at by calculations. If the design engineer has not had the experience to provide this factor of good judgment, it is always desirable to allow the company supplying the valve to check the analysis.

A method for sizing control valves that has proved to be accurate and practical is termed the C_v method and can be used for all flowing fluids. By definition, the valve flow coefficient C_v is "the number of gpm of water which will pass through a given flow restriction with a pressure drop of 1 psi." All valve manufacturers include in their literature formulas for converting the basic C_v to steam, gas, or vapors.

Flow Control

All flow-control problems require valves having the proper flow characteristics, and the valve theory just covered on flow-valve characteristics will have its widest application here because in problems of flow control, the system usually has hardly any demand side-storage capacity. If an orifice is placed in a pipeline and a controller installed to control the flow of fluid through this line, it is apparent that there is practically no storage capacity in the system, and any sudden change in demand must be corrected immediately by the control mechanism. This action presents the most difficult control problem, and any characteristic of the control to induce cycling must be avoided. Hence, in flow control the controller mechanism must almost always be the proportional-and-reset type, and the control valves must have the best possible throttling characteristics. Simple controllers and ordinary valves not designed for flow control usually result in complete failure of the control.

Control of Flow by Throttling against a Constant Pressure

This is the simplest type of flow control and is used for both liquids and gases. The first system to be considered will be that of liquid control.

Liquid Control

Where a control is installed in a system that is under a constant pressure maintained by means of a second control or where the pressure is hydrostatic, the simple arrangement of controller shown in Fig. 9-10 works satisfactorily. In this case the valve is placed downstream from the orifice plate.

If a fluid is to be controlled by throttling against a pump pressure, provision must be made to prevent destructive pressures from being built up in the pump when the valve is closed. This is necessary only where delivery is made through a positive-displacement type of pump. Figure 9-11 shows the arrangement of a positive-displacement pump equipped with a flowmeter and controller having the control valve installed in a bypass line returning to the suction side of the pump. In this case any change in flow results in an adjustment of the amount of fluid bypassed and a constant delivery maintained in this way.

If the delivery is by means of a centrifugal type of pump, it is possible to throttle against the pump pressure without building up excessive

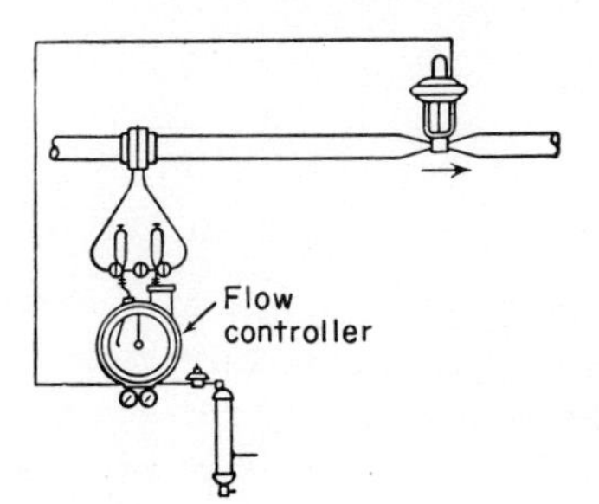

Fig. 9-10 Constant-pressure flow controller.

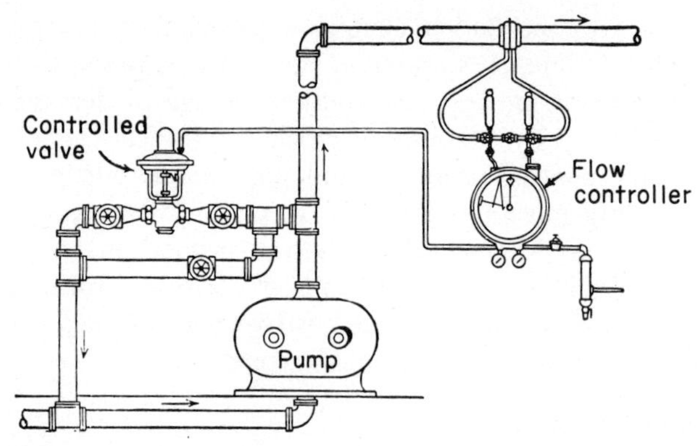

Fig. 9-11 Flow control of positive-displacement pump.

pressure. This being the case, it is possible to use the simple system shown in Fig. 9-12. This diagram shows the control with a bypass valve which may be used to carry part of the flow. However, for best control the entire flow should pass through the control valve, using the bypass only in case of failure of the control mechanism.

Control of Gas Volume

The control of gas flow is similar to that of liquid flow, except for the problems resulting from the compressibility of the gas. The position of the control valve depends upon whether the flow is to be controlled on the basis of the downstream or the upstream pressure of the gas. If the downstream pressure of the gas is constant at a constant flow, the valve should be placed upstream as shown in Fig. 9-13. If the upstream pressure is constant and the downstream variable, installation is made as shown in Fig. 9-14. In either case the volume delivered is propor-

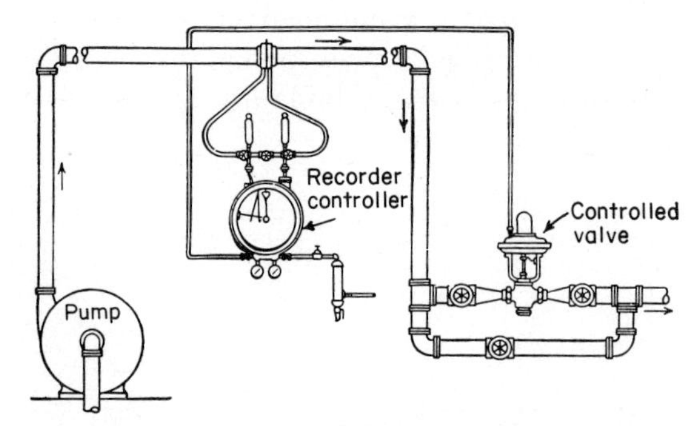

Fig. 9-12 Flow control with centrifugal pump and bypass valve.

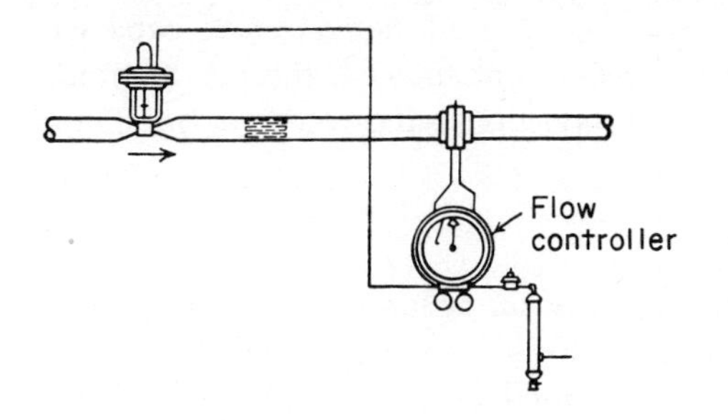

Fig. 9-13 Flow control of a gas with downstream pressure constant.

tional to the square root of the pressure at the orifice for any setting of the control instrument.

As we have seen in the chapter on flowmeters, the gas flowmeter can be compensated for static head, and with this type of meter built into the controller the control can be made on the basis of absolute volume, irrespective of variations in pressure. Installation of this type of controller is shown in Fig. 9-15.

PUMP-SPEED THROTTLING

In many cases it is desirable to control the flow of a fluid by varying the speed of the pump rather than by throttling the flow in the discharge line. This is done in many cases to avoid the expense of a control valve large enough to throttle against large volumes of flow, and other means must be provided.

Flow-ratio Control

Flow-ratio control is obtained by measuring the flow of the two or more media to be proportioned and then controlling the secondary flow or flows in a fixed ratio with that of the primary flow. If the primary flow is in turn stabilized at a certain value, the system delivers different fluids in fixed volumes, each being a definite proportion of the total flow. This type of control is difficult for the same reasons that

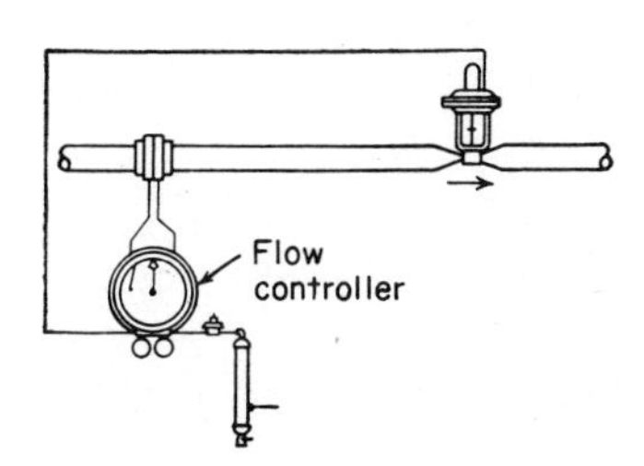

Fig. 9-14 Flow control of a gas with a constant upstream pressure.

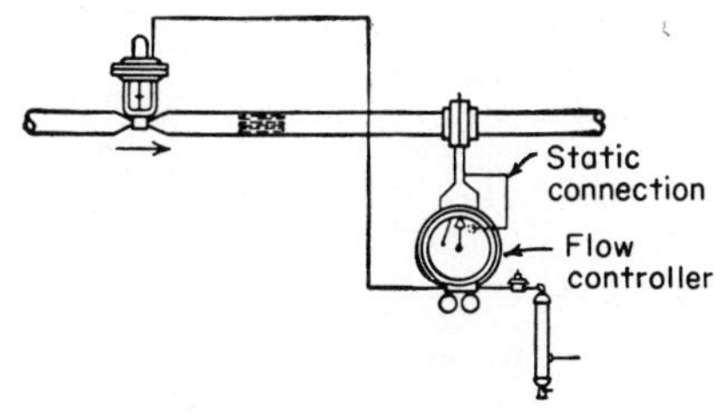

Fig. 9-15 Flow control of a gas with static-head compensated controller.

simple fluid flow is difficult. There is little or no demand-side capacity in the system, and if the two flows become unbalanced, there is a strong tendency for the system to cycle. Proportional-and-reset control must always be used.

Single-instrument Flow Control

The use of a flow-control mechanism of an absorber is shown in Fig. 9-16. Here the wet gas is the uncontrolled variable, and the oil is the controlled variable which is maintained at a constant ratio to the gas. Both the flows are measured and balanced in the meter in such a way that they establish equilibrium at the point of proper ratio. If the flows become unbalanced, a differential force is set up in the pneumatic system which repositions the control valve to reestablish the correct flow ratio. Note that there is only one proportional-and-reset-control mechanism that operates from the differential responses set up in the meter. An adjustment is provided that changes the point of balance and in this way permits the flow ratio to be changed at will.

Deviations in the primary flow up to 75 percent or more can be con-

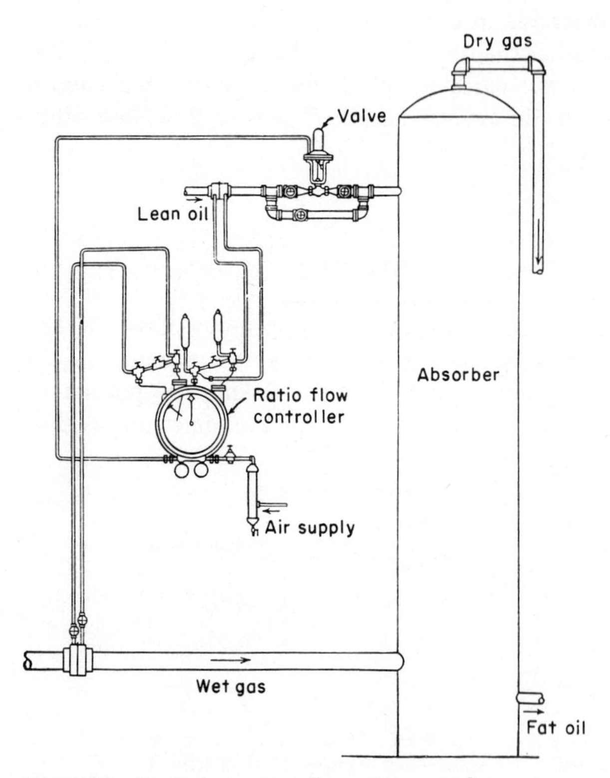

Fig. 9-16 Single-instrument flow-ratio control.

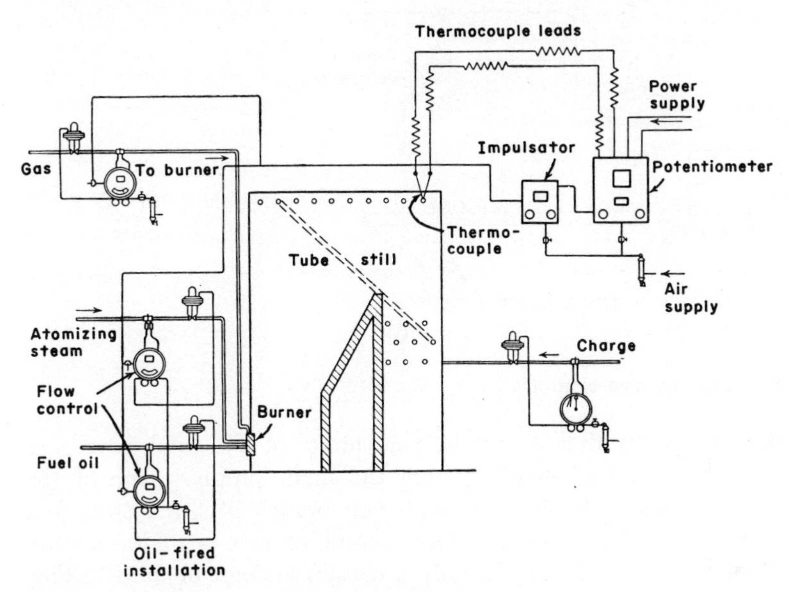

Fig. 9-17 Flow-ratio control of a pipe still.

trolled in correct ratio with a setup of this type. Greater deviations in flow are likely to cause trouble as a result of the orifice characteristics being different over extremely wide ranges of flow.

Two-instrument Flow Control

If the system of flows to be controlled is very unstable because of large lags in the process and secondary variations in pressure, etc., it is often necessary to provide separate proportional-and-floating-control units on each of the flows to be controlled. This also can be done in the case of unstable liquid-level control. With this system of ratio control each flow is set for the correct adjustment of throttling range and reset adjustment, and the set point of each is varied in exact ratio with the flow measured by the primary controlling instrument. A system of this type is shown in Fig. 9-17 describing a pipe-still control. Here a potentiometer is the primary control device, and the flows of fluid oil, steam, and gas to the furnace are the controlled flows. Note that the potentiometer controls the set point of the fuel oil, the steam, and the gas so that each varies simultaneously with changes in temperature.

Split-feed Control

A further use of two-instrument flow controls is that shown in Fig. 9-18. Here a flow is proportioned automatically by means of a three-

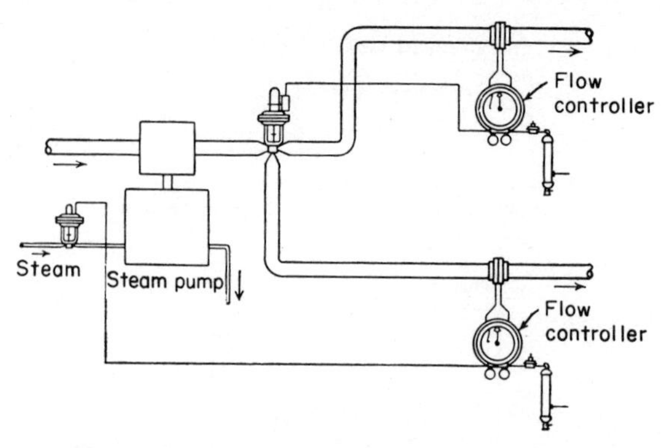

Fig. 9-18 Split-feed flow control.

way valve to ensure that a certain percentage of the total flow goes to each of the two lines leading from the main supply. One of the instruments measures the flow through one branch of the system and controls the flow to the pump. The second branch is under control of the second meter and permits only a certain amount of fluid to flow through this pipe. The remainder flows through the first line, and if too much or too little is available, the pump speed is automatically changed. As a result the entire system is in balance. Both of these controllers must be the proportional-and-reset type in order to prevent serious cycling of the system.

Time-flow Control

It is often desirable to vary the flow of materials to a process on a certain definite time schedule. This can be done simply by means of a clock-driven cam which varies the set point of the controller according to a set time schedule.

The illustrations used in this section show an early model of The Foxboro Company Stabilog* control in use.

* Registered trade name, The Foxboro Company.

CHAPTER TEN

Liquid-level and Interface Measurement and Control

MEASUREMENT OF LIQUID LEVEL is one of the three fundamental measurements most frequently required in a chemical process; the other two are temperature and flow. Measurement of the liquid level in a vessel may be made simply as a check on the supply of material on hand as a means of determining the exact amount of liquid to be delivered to a batch process or as the primary measurement in a control system designed to maintain a certain level in a vessel that is part of a continuous process. The latter use for liquid-level measurement is becoming increasingly important in the chemical industries, especially in such industries as petroleum refining where continuous distillation processes are widely used. Liquid-level-measuring instruments are also widely used in waterworks and hydroelectric- and steam-power plants, etc.

Float-displacement-type Element

BLACK, SIVALLS & BRYSON, INCORPORATED

PRINCIPLE OF DESIGN: Figure 10-1 is a view of the float, float cage, and torque tube of a tubular-float-displacement-type liquid-level con-

* Grady C. Carroll, *Industrial Instrument Servicing Handbook*, 1st ed., McGraw-Hill Book Company, New York, 1960, pp. 4-9, 4-10.

troller of the Black, Sivalls & Bryson, Incorporated, design. Float cages are available with various connection combinations for side and bottom (as shown), top and bottom, and side connections. The principle upon which the displacement-type liquid-level controller operates is based on the fact that an object, which is the tubular-displacement element, weighs less when submerged in a liquid than when in air or other lighter fluids. This element does not actually float when in operation.

The heart of the controller is the torsion tube, referred to in many applications of similar design as a torque tube. The torsion element consists of a metal tube with one end firmly held to the torsion-tube housing by a plate, the other end being free to turn. A torsion arm extends at right angles from the free end of the tube onto which is suspended the displacement element. The weight of this element is transmitted to the torsion tube through the leverage of the arm and causes a twist in the tube which varies in proportion to the weight imposed by the rising and lowering of the liquid in the displacer cage. As the liquid rises, the element is more fully submerged, thereby reducing the weight at the end of the arm.

Within the torsion tube, there is a ⅛-in. rod which is welded inside the free end of the torque tube and extends outside the fixed end of

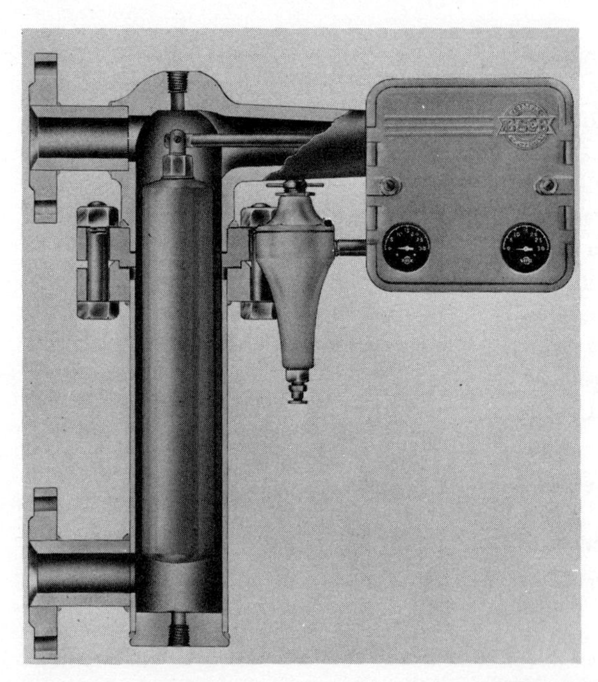

Fig. 10-1 Tubular-float-displacement-type liquid-level controller. (*Black, Sivalls and Bryson, Inc.*)

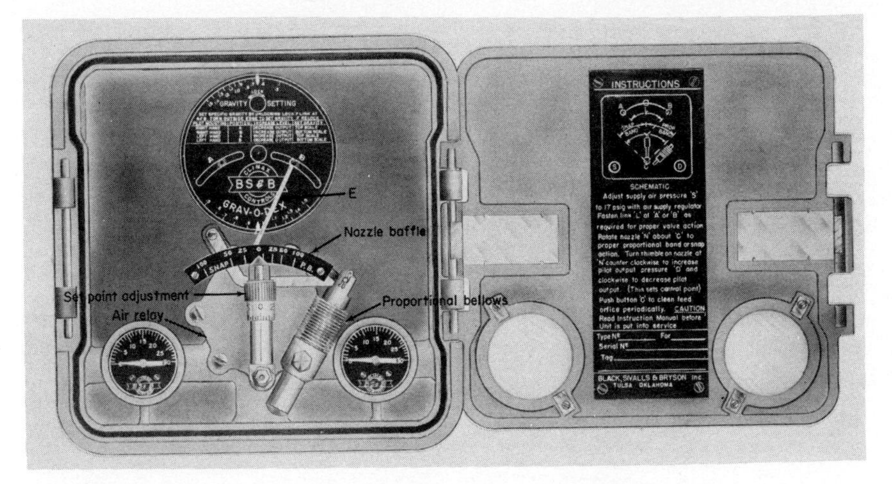

Fig. 10-2 Pneumatic controller. (*Black, Sivalls and Bryson, Inc.*)

the tube through a set of ball bearings into the controller case shown in Fig. 10-2. Within the case is mounted the control mechanism, the nozzle baffle of which is connected through a link to the torsion-tube shaft. Thus the exact amount of turning of the torsion tube, resulting from the weight of the displacer, is transmitted to the control baffle. Between the torsion-tube shaft and the baffle is imposed a unit for correcting the torque to the actual specific gravity of the liquid in contact with the displacer, as noted on the Grav-O-Dex scale. Instructions are provided on the disk to assist in properly setting the gravity unit.

The vertical movement of the displacer, caused by displacement of liquid, is transmitted through the torsion tube, Grav-O-Dex unit, and link *E* to the nozzle baffle. The position of the baffle in relation to the bleed nozzle determines the back pressure of the nozzle. This back pressure is applied through drilled passages to the pilot-valve diaphragm, which produces a pilot-valve output pressure to the control valve that is proportional to the nozzle back pressure. Thus for every position of the displacer, there is a proportional output pressure applied to the control-valve diaphragm, the value of which depends upon the proportional-band setting of the controller. The "proportional band" is the percentage of total vertical travel of the tubular displacer used for a change in pressure output from the controller of 3 to 15 psi. Variations in the proportional band are made possible by movement of the bellows unit.

The Black, Sivalls & Bryson displacement-type liquid-level controller shown in Fig. 10-1 is similar to Fisher Governor Company's Level-Trol.*

* Registered trade name, Fisher Governor Company.

Fisher Governor Company*

PRINCIPLE OF DESIGN: The float-displacement-type level-measuring instrument is based on the apparent change in weight of a body when it is placed in a liquid. According to Archimedes' principle, a body placed in a liquid is buoyed up by a force equal to the weight of the displaced liquid. All float-displacement-level instruments use this principle, the major difference being in the methods used to weigh a float and convert its weight into liquid-level measurement.

Shown schematically in Fig. 10-3 is a Fisher Governor Company displacement-type instrument which uses a torque tube as a weighing de-

* Grady C. Carroll, *Industrial Process Measuring Instruments,* 1st ed., McGraw-Hill Book Company, New York, 1960, pp. 165–167.

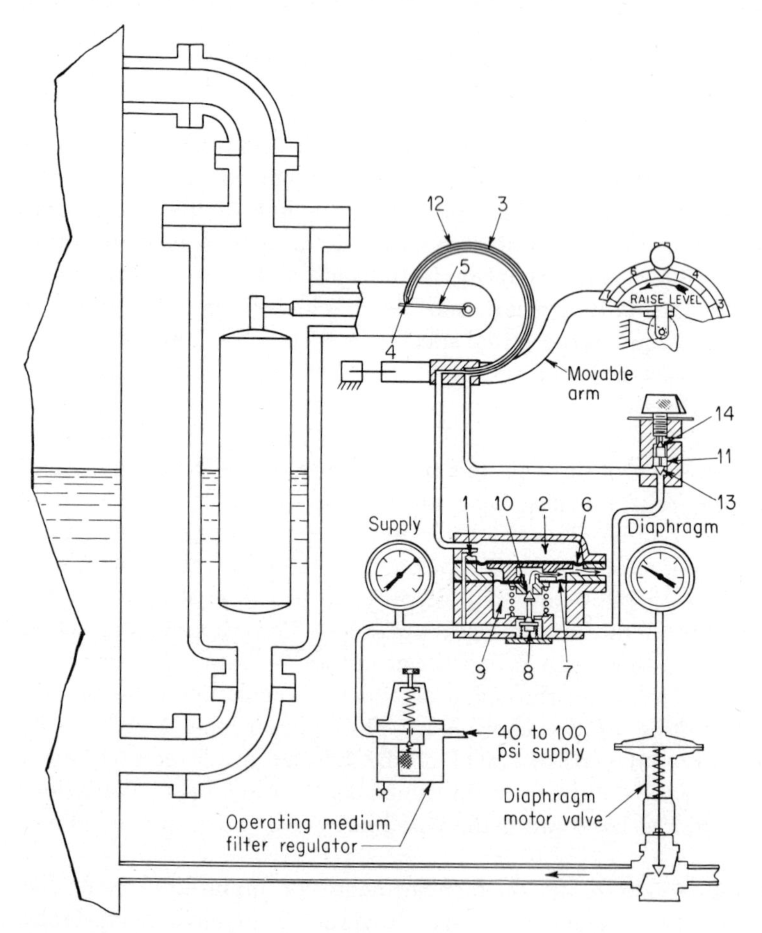

Fig. 10-3 Schematic of a displacement-type liquid-level controller. (*Fisher Governor Company*)

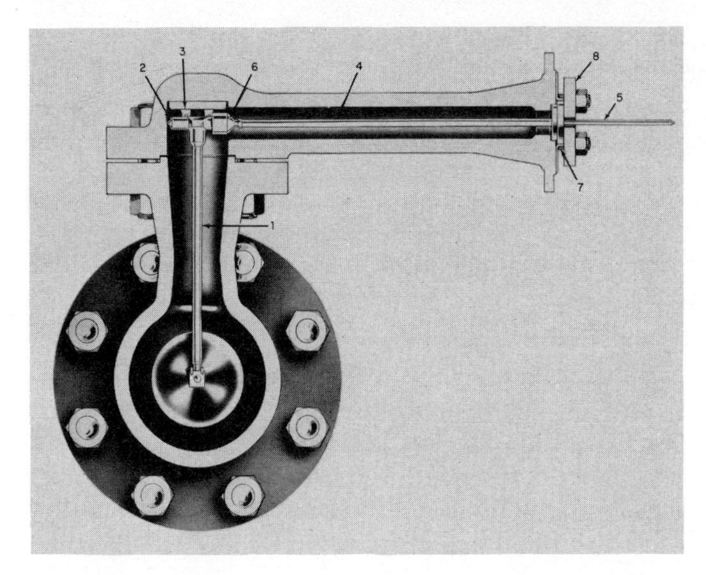

Fig. 10-4 Torque-tube assembly. (*Fisher Governor Company*)

vice to detect apparent weight changes of the float as the liquid whose level is being measured rises and falls around it. To understand better the function of a torque tube, refer to Fig. 10-4, which shows the float and torque tube of a Fisher Governor Company Level-Trol. The torque tube consists of two parts: metal tube 4 with shaft 5 inside it. The two are welded together at 6. A flange is welded to the other end

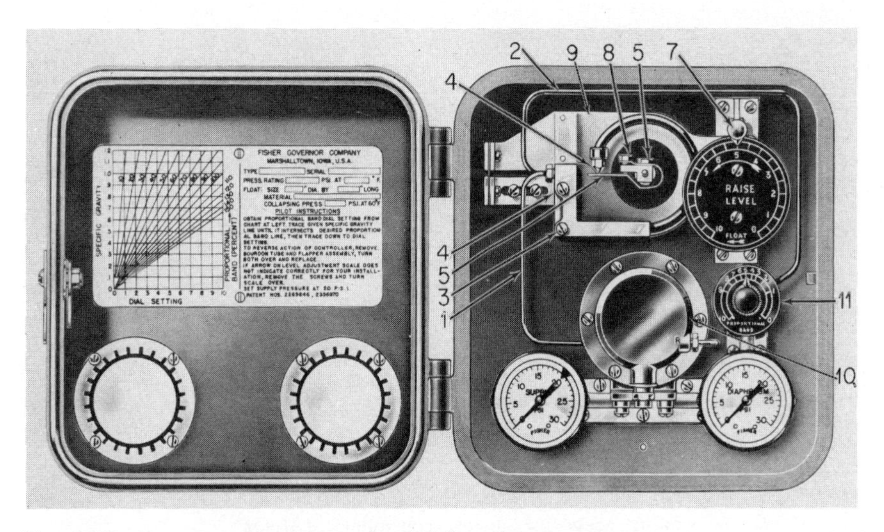

Fig. 10-5 Pneumatic level control. (*Fisher Governor Company*)

of the tube and is securely clamped between flanges 7 and 8. The shaft extends into the instrument case shown in Fig. 10-5, where a flapper-and-nozzle system converts its rotary motion into a 3- to 15-psi pneumatic signal which represents a level change proportional to the float weight.

MOORE PRODUCTS CO.*

PRINCIPLE OF DESIGN: As can be seen from the schematic in Fig. 10-6, Moore Products Co. uses an adjustable spring without a torque tube

* Carroll, *Industrial Process Measuring Instruments*, pp. 167–172.

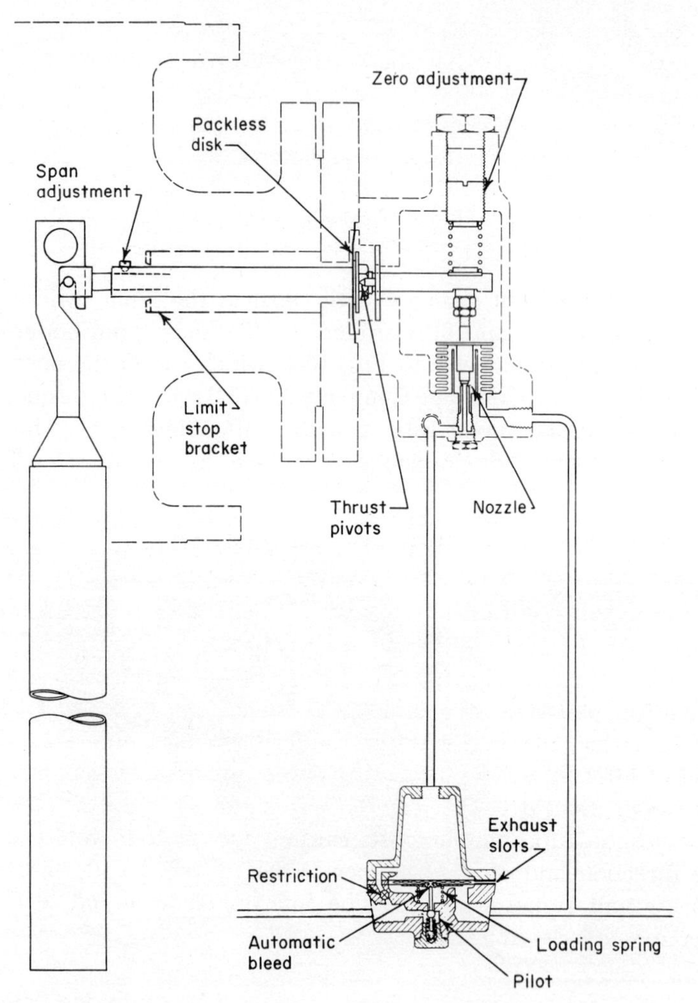

Fig. 10-6 Schematic view of level controller. (*Moore Products Co.*)

to balance the weight of the float at zero position when no liquid is in contact with the float. The bellows-and-pilot assembly is used to convert the buoyant force of the float into a pneumatic output signal of 3 to 15 psi. Note that a diaphragm is used as a pressure seal for the float shaft rather than a torque tube or packing gland.

The float-displacement-type level-measuring instrument is designed for pneumatic transmission in most cases. The method used by the majority of manufacturers for converting float movement into a pneumatic signal is some form of the conventional flapper-and-nozzle system with an amplifying pilot valve. However, Moore Products Co. uses a null balance system consisting of a bellows-and-pilot valve without a flapper and nozzle. Either method is rugged and reliable.

There are two reasons for the popularity of the pneumatic transmitting system: (1) the general acceptance of the pneumatic system for process control prior to the development of the electronic system and (2) the relatively slow process response usually encountered in liquid-level measurement and control, which reduces the need for high-speed transmission to a minimum. Another reason of less importance is the simplicity of the pneumatic transmitting and receiving instruments, which makes them desirable from a maintenance point of view, especially in small plants where a minimum of skilled labor is available.

Transmitting, Pneumatic

The torque tube is designed so that the weight of the float with no liquid in contact with it twists the tube a maximum amount in one direction, thereby rotating the shaft in such a way that it produces a 3-psi pneumatic output from the transmitter, which is equal to zero level as far as the float is concerned since no liquid is in contact with it. Refer to Fig. 10-4. If liquid is allowed to enter the float cage until 4 in. of the float is immersed in it, the float loses weight equal to the weight of liquid displaced by it (see Figs. 10-3 and 10-4). If the float is 3 in. in diameter and the fluid is water, the float will have an apparent loss of weight of 1.071 lb, which means that the torque tube is supporting 1.071 lb less weight than it was before the water entered the float cage. Therefore the torque tube will untwist, causing the shaft to rotate in the opposite direction and operate flapper 5, thereby producing an increase in the output signal which can be interpreted as liquid level. This is proved by the following formulas:

$3 \times 3 \text{ in.} \times 0.7854 \times 4 \text{ in.} = 29.594$ cu in. of water displaced by the float

$29.594 \times 0.0362 = 1.071$ lb of water displaced by the float

where 3 in. = diameter of float
4 in. = depth of liquid in float cage
0.7854 = factor
0.0362 = weight of 1 cu in. of water, lb

In the above case the output signal would have increased from 3 to 6.42 psi if the float were 14 in. long.

Electrical-contact-type Indicators

The Vapor Recovery Systems Company*

PRINCIPLE OF DESIGN: The float-and-tape gauge shown in Fig. 10-7 is not designed for pneumatic transmission; gauges of this type transmit by electrical rather than pneumatic means.

For long-distance transmission the float-and-tape liquid-level gauge uses some type of electrical system.

The Vapor Recovery Systems Company gauge shown in Fig. 10-7 uses a transmitting unit which operates on the coded-electrical-pulse principle. By an arrangement of dots and dashes, for each eighth increment of liquid level from 0 to 59 ft 11⅞ in. or greater, there exists a separate and distinct code. Each code group, as transmitted from the tank unit, is the correct liquid-level value, which is produced by the receiver-indicator. Each code group is an entity in itself; partial

* *Ibid.*, pp. 163–164.

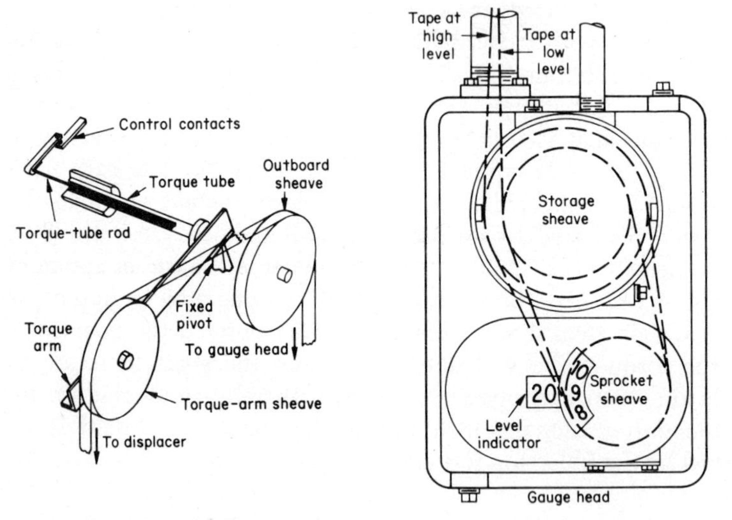

Fig. 10-7 Electrical-type liquid-level transmitter. (*The Vapor Recovery Systems Company*)

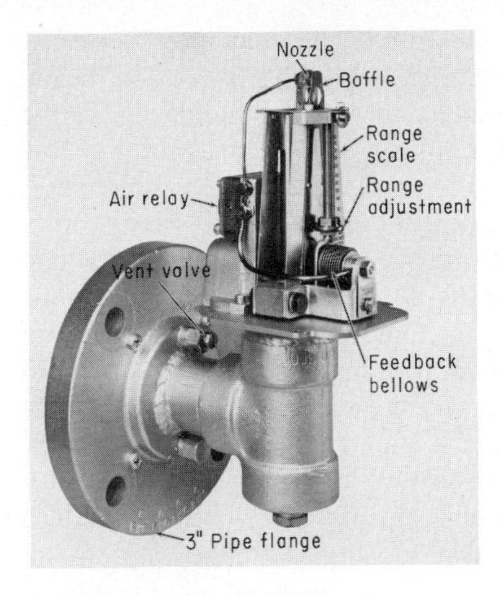

Fig. 10-8 Pneumatic-type liquid-level transmitter. (*The Foxboro Company*)

codes or grouping of codes is eliminated. The loss or gain of any part of the code produces a signal that has no meaning to the receiver, and as a result no indication appears. Thus the receiver ignores any false signal. The transmitting unit can be enclosed in an explosion-proof fitting for hazardous locations.

A telemetering system is available whereby almost any number of tank levels can be read from one station. This consists of a slow-speed coded pulse with a dial similar to a telephone dial for selecting the pulse codes which correspond to tanks. The level in any tank in a particular system can be read by dialing its code. Transmission can be made over one-wire-and-ground, two-wire, or any type of carrier system operated over existing telephone lines or microwave.

The Foxboro Company

Figure 10-8 is a Foxboro Company pneumatic transmitter designed specifically for liquid-level measurement. The transmitter can be used on pressure or vented tanks. It is mounted on a 3-in. flange attached to the vessel in which the level measurement is being made. The stainless-steel diaphragm is in contact with the measured material at all times. The conventional Foxboro transmitting unit is mounted on the 3-in. flange and operates to balance the measured liquid head by the transmitted pressure, which can be transmitted several hundred feet to a receiving or recording instrument and recorded or indicated in the proper units of measurement.

The transmitter connecting pipe is usually attached to the vessel about 2 in. above the bottom to avoid the trapping of sediment in the transmitter body. If this is not done, the instrument may need frequent cleaning to maintain its accuracy, particularly if it is located below the tank bottom. In this case the sediment or dirt builds up in the connecting pipe and changes the calibration.

Air- or Inert-gas-bubble Type of Liquid-level Instrument

Because of the possibility of losing the air in the air-trap type of pressure bulb as a result of absorption by the liquid or by condensate forming in the tubing, it is often desirable to use an auxiliary air supply to assure that the liquid-air surface is at the base of the tank. In this arrangement, air is forced into the manometer or recording instrument at a point along the tubing leading from the instrument to the pressure bulb, which is of the open type without diaphragm. This air bubbles out of the pressure bulb and prevents the liquid from rising in the tube. Regardless of the pressure existing at the point of air supply, the pressure at the instrument is that caused by the static pressure of the liquid because if the air pressure exceeds this value, it escapes from the system and bubbles to the surface of the vessel. This system permits the indicator or recorder to be located at a considerable distance from the vessel. In the case of the simple air-trap instrument or the instrument using the diaphragm-sealed pressure bulb, it is necessary to build the air trap large enough to provide sufficient air to prevent the tube system from filling with liquid at the maximum static pressure. If the connecting tubing is long and the static pressure high, the tubing must be small in diameter to reduce its capacity. This in turn multiplies the danger of the tube system's being clogged. With the air-bubble system the connecting tube may be larger, and the escaping air tends to keep it clean.

Several arrangements of the air-bubble type of liquid-level indicator

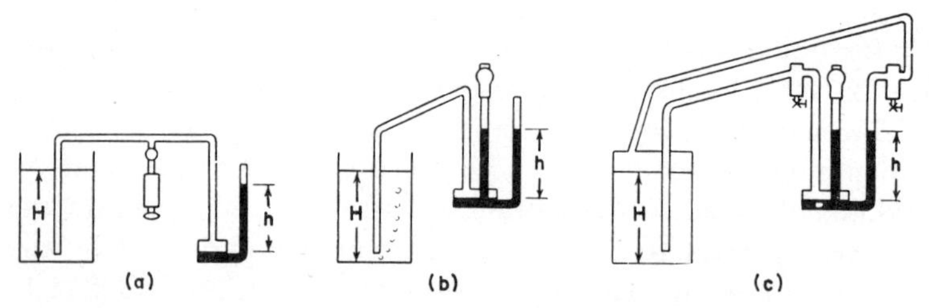

Fig. 10-8a Simple well-type manometer liquid-level indicator with hand pump.
Fig. 10-8b Manometer liquid-level indicator with aspirator.
Fig. 10-8c Manometer liquid-level indicator with aspirator for pressure vessel.

are shown. Figure 10-8*a* shows a simple well type of manometer with a hand pump to supply air to the system. When a reading is taken, the operator pumps air into the system until the mercury in the column stops rising. The manometer then indicates the static pressure at the base of the tank or vessel. Figure 10-8*b* shows a well manometer with an aspirator to provide the air pressure for operation. In this design the mercury column is connected to the air-delivery column to prevent a sudden surge of pressure from blowing the mercury out of the manometer. A third arrangement is shown in Fig. 10-8*c*, where the manometer column is connected to the top of the vessel to permit operation with the vessel under pressure. Traps are provided if the fluid is volatile. Any of these three manometer arrangements can be used with a slanting manometer where greater sensitivity is required.

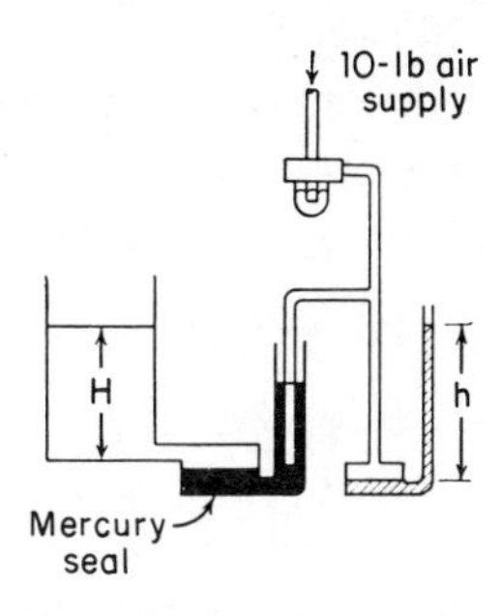

Fig. 10-9 Manometer liquid-level indicator with mercury seal-air-bubble system.

If it is undesirable to have the air bubble through the liquid being measured, an arrangement similar to that shown in Fig. 10-9 can be used. Here the liquid in the vessel causes the mercury in the mercury-sealing manometer to rise. A pipe is extended into the mercury column, and an air supply bleeds air into the secondary manometer system connected to it. As a result the liquid level can be read directly on the secondary manometer. Note that a bubble chamber is used as a trap in the air-supply line. This feature is used in nearly all cases where a continuous flow of air is used to measure liquid level and provides a means of trapping dirt and foreign material that might clog the manometer tubing and also gives a visual check on the rate of airflow into the system. If a large volume of air were permitted to move through the system, the resistance of the pipe to the airflow would build up a pressure in the indicator that would be higher than the true static pressure. Only a small flow of air is necessary to maintain the liquid surface at the bottom of the tank and to assure the operator that the piping is free of obstruction.

Pressure-differential-type Instrument

In most cases where the liquid-level measurement is made primarily to give direction to a controlling mechanism, the range of liquid-level movement is comparatively small. When a liquid-level water-pressure head of 200 in. or less is developed in the vessel, it is possible to use a standard recording- or indicating-flowmeter manometer to measure the liquid level. This is advantageous not only because it permits the use of an instrument that is produced in large quantities and consequently at a low price but also because these instruments are available for any working pressure.

Purge Systems*

In processes where the level of extremely viscous materials or liquids that must be prevented from contacting the measuring equipment are to be measured, liquid purges can be used as a substitute for a sealed type of instrument. However, a purge system adds equipment to the installation which must be serviced and maintained along with the measuring instrument and should be used only on installations where other methods are not practical.

In processes where liquid purges would be objectionable but air or inert gases would be tolerated, a system such as shown in Fig. 10-10 is practical, provided the purge stream can at all times be maintained

* *Ibid.*, pp. 174–176.

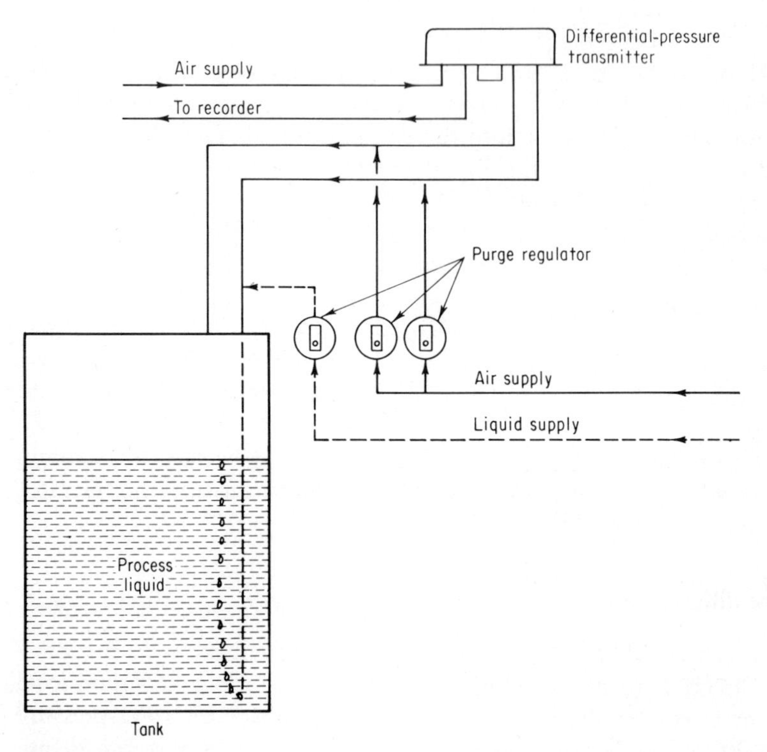

Fig. 10-10 Inert gas and liquid purge with instrument mounted above tank.

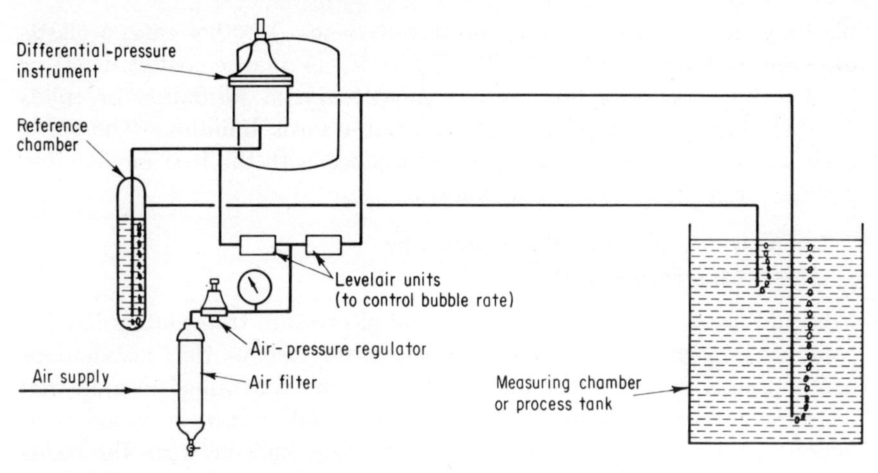

Fig. 10-11 Low differential-pressure-measuring bell-type meter. (*Grady C. Carroll*)

at a pressure exceeding that of the process material by at least 25 psig.

The long dip pipe should be securely supported inside the vessel, and its length should be at least 6 in. longer than the span over which the level is to be measured. To prevent large bubbles from forming at the end of the dip pipe and causing an unsteady output of the transmitter, two ¼-in. V's should be cut 180° apart in the end of the pipe. To eliminate any possibility of an error resulting from the purge-stream flow and for mechanical strength, the use of dip pipes having diameters less than ½ in. should be avoided. Pipes with a 1-in. diameter would be recommended for most installations.

For measuring the level of slurries and corrosive liquids in vented or open tanks, an air-purge system operating in conjunction with a differential-pressure instrument (as shown in Fig. 10-11), with the low-side pipe disconnected, makes a simple and reliable installation. With an installation of this type it is impossible for the process material to come in contact with the measuring instrument because the end of the dip pipe extends well above the top of the tank. Note that a liquid purge is also shown in Fig. 10-10 by the broken line. This would be used only where the process material in the tank formed crystal or other deposits on the end of the dip pipe as a result of the agitation caused by the flow of purge air. In such a system a small flow of some liquid is used to keep the end of the dip pipe washed clean so that the flow of purge air is not obstructed.

With some process materials, crystals and solids can be prevented from forming on a metal dip pipe by coating it with some form of plastic material such as Teflon* or polyethylene. In other cases a plastic dip pipe can be used if conditions permit. A plastic-coated metal or plastic dip pipe does not always prevent crystal formation or solids buildup, but in most process materials it retards buildup. Therefore, each installation must be made in accordance with the best engineering practice after the conditions are known.

Summary of Liquid-level Measurement by Differential-pressure Methods

Liquid-level measurement by differential-pressure transmitters has become so important to modern process-control systems that installations can no longer be left to the judgment of anyone not trained in industrial instrumentation. The application engineer should adhere to sound engineering principles, such as avoiding long gauge lines between the transmitter and process vessel in which the level is being measured.

* Registered trade name, E. I. du Pont de Nemours & Co.

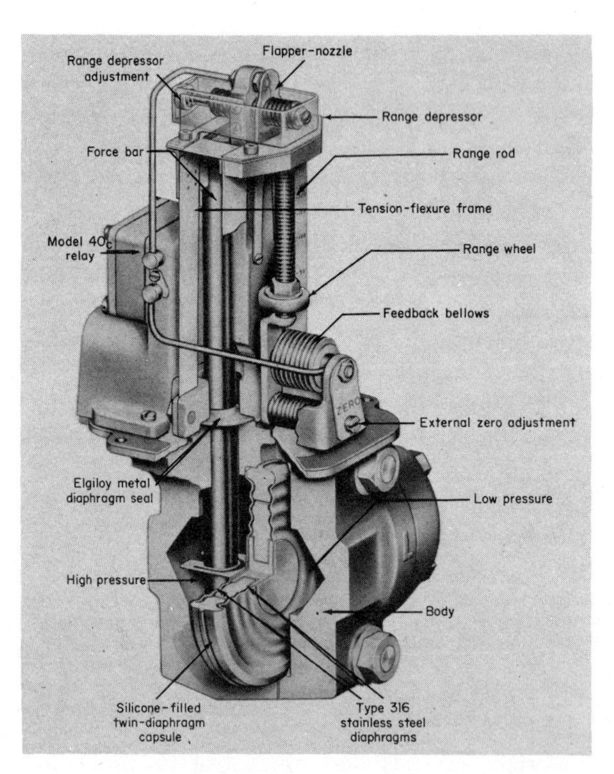

Fig. 10-12 Differential-pressure transmitter used as level-measuring instrument. (*The Foxboro Company*)

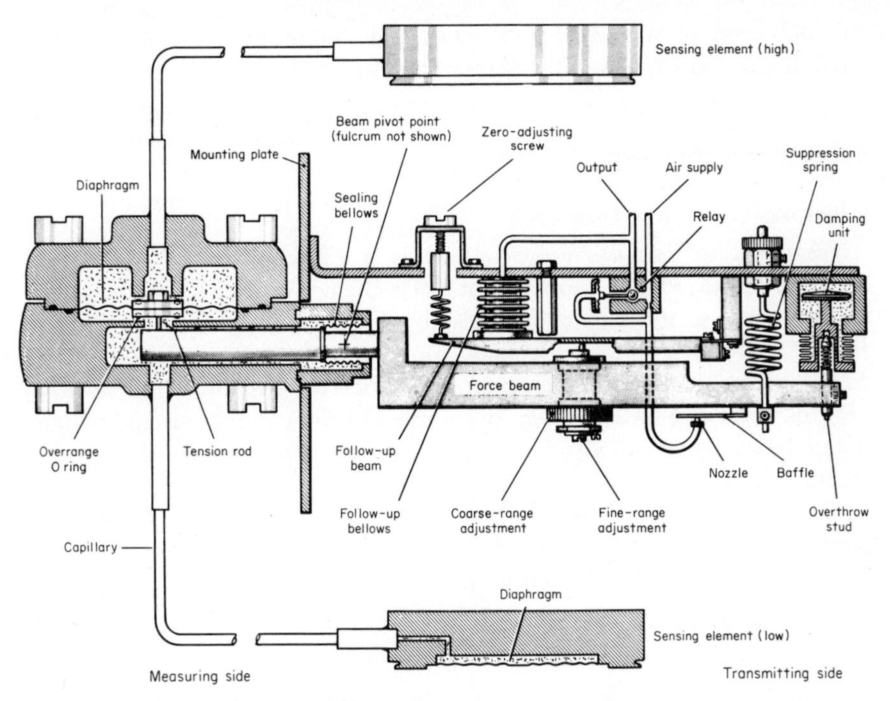

Fig. 10-13 Sealed-type differential-pressure transmitter. (*Taylor Instrument Companies*)

Force-balance, Differential-pressure-type Elements

The Foxboro Company*

PRINCIPLE OF DESIGN: In many chemical processes the measurement of liquid level in tanks and process vessels by the differential-pressure method is practical, reliable, and economical. The measuring instrument can be any of the several differential-pressure types described in previous chapters. Shown in Fig. 10-12 is one of The Foxboro Company design. The principle upon which these instruments operate is clearly set forth in other chapters, making further discussion here unnecessary.

Taylor Instrument Companies†

PRINCIPLE OF DESIGN: In processes where the liquid level of highly corrosive materials or slurries must be accurately measured, a sealed-type differential-pressure transmitter of the kind shown in Fig. 10-13 is practi-

* *Ibid.*, pp. 172–173.

† *Ibid.*

cal and economical. The instrument seal diaphragms and flanges are available in 316 stainless, Hastelloy B, and other alloys.

Capacitance-probe-type Element

Fisher Governor Company

PRINCIPLE OF DESIGN: Figure 10-14 is a Fisher Governor Company series 3100 Lev-Al-Con* instrument. The Fisher series 3100 Lev-Al-Con instruments use the capacitance-sensing principle to provide control of liquid or granular-solid level. The complete series consists of type 3105 for single-point level alarm, type 3110 for two-point differential-level control, and type 3120 for proportional-level-control or indication. These instruments can also be used for fluid-interface detection (nonconducting fluids).

The sensing element is the type 312 capacitance probe. Instrument set-point adjustments shown in Fig. 10-15 allow the control point(s) to be set anywhere over the entire probe length.

* Registered trade name, Fisher Governor Company.

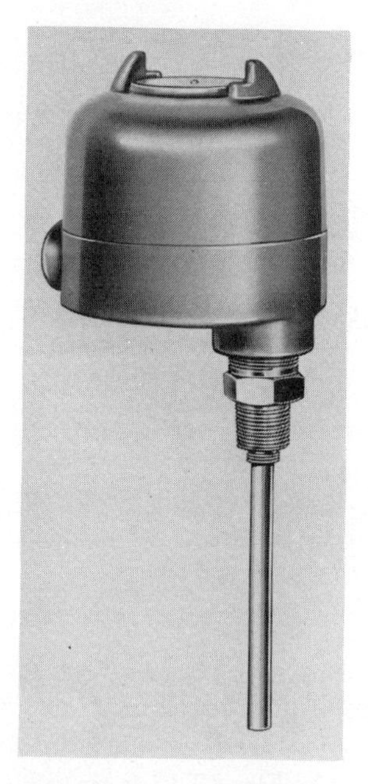

Fig. 10-14 Capacitance-probe-type, liquid-level element. (*Fisher Governor Company*)

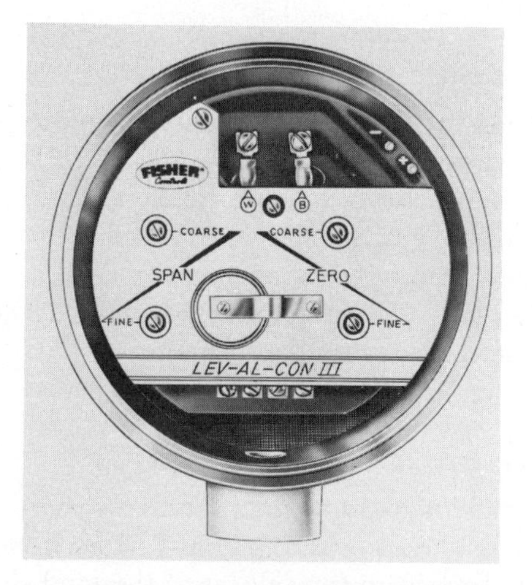

Fig. 10-15 Set-point adjustment for capacitance probe. (*Fisher Governor Company*)

Series 3100 features include rugged, self-contained construction; fail-safe, high- or low-level operation; fully temperature-compensated, solid-state circuitry; plug-in chassis; explosion-proof aluminum housing; excellent sensitivity; and extreme ease of calibration.

Dielectric Constant Requirements

The dielectric constant of a material is the ratio of the capacitance of a capacitor with the material as the dielectric to the capacitance of the same capacitor with air as the dielectric (air has a dielectric constant of 1).

Any series 3100 instrument together with a sheathed probe can be used on level-control applications where the liquid has an effective dielectric constant of 1.2 or greater. With an unsheathed probe, the recommended dielectric constant range is 1.7 or greater.

For interface detection, a minimum of 0.2 difference in dielectric constants is the recommended limit for an instrument with a sheathed probe, while a minimum of 0.7 is recommended for an unsheathed probe.

The type 3105 provides single-point detection. Any capacitance between 10 pF can be selected as the alarm point and "set" into the instrument by the coarse and fine adjustments. When the capacitance of the probe (determined by material level) equals the set-point capacitance, the relay actuates.

The type 3110 provides two-point differential control. Coarse and fine set-point adjustments allow a lower control point capacitance to be chosen anywhere within a 10- to 530-pF range. Coarse and fine differential adjustments allow choosing a differential capacitance from 1 to 530 pF. The algebraic sum of these capacitances is the upper control-point capacitance.

The type 3120 senses and converts a capacitance change to a proportional milliampere signal which is then transmitted to control or alarm equipment. Two signal ranges are available: 4 to 20 ma dc or 10 to 50 ma dc.

The type 3120 can be used as either a direct-acting unit where rising level gives an increasing output signal or as a reverse-acting unit where rising level gives a decreasing output signal. Action can easily be changed from direct to reverse by having the two leads located on the top board of the transmitter interchanged.

Calibration of type 3120 is extremely easy since there is no interaction between the adjustments of the Zero and Span controls. A standard voltmeter may be connected to the calibration terminals on the top of the chassis without the load circuit being broken. The voltmeter will read 0.3 volt at the low point of the output range and 1.5 volts at the high point for both 4- to 20-ma and 10- to 50-ma ranges.

Principle of Operation

Types 3105 and 3110 have an internal power supply and a voltage regulator which converts a nominal input of 120 volts ac to a regulated dc voltage. This regulated voltage drives the instrument circuitry, with the exception of the load relay. A feedback oscillator generates a 200-kHz signal which is fed to a pair of switching transistors. These transistors together with a diode network alternately charge and discharge two capacitors. One capacitor is the probe; the other is the set-point capacitance. The capacitors discharge through the diode network with the algebraic sum of their discharge currents being the network output. This output is applied to a differential amplifier, which in turn drives the load relay. This relay has both Normally Open and Normally Closed contacts for connection to external alarm and/or control equipment, and its action is determined by the position of the Fail Safe switch.

FAIL-SAFE OPERATION: A switch located on the face of types 3105 and 3110 chassis permits either fail-safe, high-level (FSHL) or fail-safe, low-level (FSLL) operation. This gives the option of having the load relay deenergized on either high or low level.

When type 3105 is operating in the FSHL mode, the relay is energized if probe capacitance is below the operating-point capacitance. Coarse- and fine-operating-point adjustments allow an operating capacitance to be chosen anywhere within a 10- to 530-pF range. It is deenergized when it is above the operating-point capacitance. When operating in the FSLL mode, the relay is deenergized when probe capacitance is below the operating-point capacitance and energized when probe capacitance is above it.

In type 3110 when probe capacitance is less than the lower control-point capacitance, the load relay is energized in the FSHL mode. As probe capacitance exceeds that of the upper control point (sum of lower control point and differential capacitances) the load relay is deenergized, and a second relay, which is controlled by the load relay, switches the differential capacitance out of the circuit. This action makes the lower control point the load-relay "trip" point. The load relay is again energized, and the differential is switched back into the circuit when probe capacitance becomes less than that of the lower control point.

The energized state of the load relay is reversed when operating in the FSLL mode. The action of the differential relay is the same regardless of the Fail Safe switch's operating mode.

Figure 10-16 is a Fisher Governor Company type 3120-312C liquid-level transmitter mounted in a Fisher Governor Company conventional liquid-level cage. The type 3120 has an internal power supply and voltage regulator requiring a nominal input of 120 volts ac. The power supply and regulator convert the ac input to the dc voltages necessary for the rest of the instrument.

A crystal-controlled oscillator generates a 200-kHz signal for use in the switching circuit. The switching circuit produces a square-wave signal that alternately charges and discharges two capacitors. One capacitor is the probe; the other is the zero adjustments. These capacitors discharge into the detecting network, with the algebraic sum of their capacitance-discharge currents being the network output current. This current is the summing point input to the dc amplifier. Negative feedback from the output dc amplifier is also introduced at this summing point and controls the "gain" or span of the instrument. The summation of these currents is the drive to the dc amplifier. The dc amplifier converts this current to the output range.

By adjustment of both Zero controls, any capacitance between 10 and 530 pF can be set into the zero side of the instrument. When the output current is at the zero output point, either 4 or 10 ma, the capacitance of the zero adjustments is equal to the total capacitance that the instrument "senses" on the probe side of the instrument.

TYPE 312 CAPACITANCE PROBE: The type 312 capacitance probe used with the Lev-Al-Con series is available in the following styles:

Style LT: insulated low-gain probe
Style MT: insulated medium-gain probe
Style HT: insulated high-gain probe
Style B: uninsulated bare-rod probe

All these probe styles can be furnished with a stainless-steel sheath

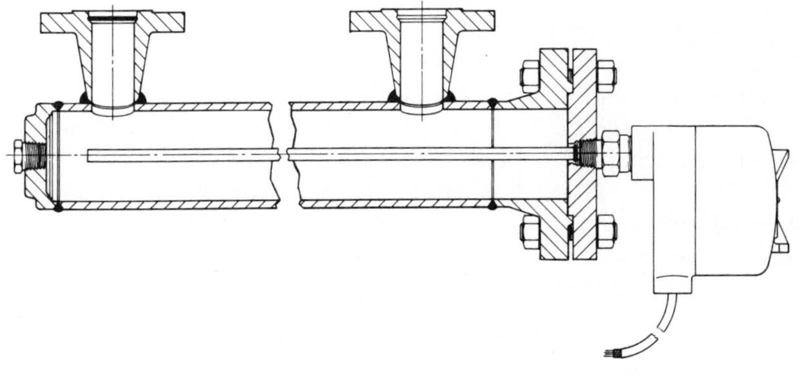

Fig. 10-16 Capacitance probe mounted in conventional liquid-level cage. (*Fisher Governor Company*)

extending the full length of the probe. In this case, the probe style becomes LTS, MTS, HTS, or BS.

The type 312 probe is available with either a ¾- or 1-in. NPT mounting connection and may be ordered in lengths of 6 to 72 in. in 6-in. increments.

The type 312C probe is the type 312 installed in a cage similar to standard Level-Trol displacer cages.

REMOTE-PROBE MOUNTING: A remote-probe mounting assembly allows separate installation of the series 3100 instrument and the type 312 probe. The probe is mounted on the vessel and connected by electrical cable to the instrument; the maximum standard cable length is 42 ft. A capped elbow protects the probe-cable connectors; a special instrument base accepts the other cable end. A special circuit chassis is also required.

Liquid Interface

Tubular-float-displacement Type of Element

Fisher Governor Company*

PRINCIPLE OF DESIGN: The Fisher Governor Company specific-gravity-measuring instrument discussed at the beginning of Chapter 8 can also be used for measuring the interface between two immiscible liquids such as water and gasoline.

There are two factors which must be considered when specifying a float-displacement-type instrument for interface measurement: the float size and the minimum, acceptable, proportional-band setting.

To determine the size of float required for any interface-level-measuring application, the following information must be known:

1. Length of displacer required
2. Specific gravity of the two liquids
3. Minimum proportional band (in inches) desired
4. Style of Level-Trol to be used, i.e., cage, top internal mounted or side internal mounted.

With the above information given, the float size for any interface-level-measurement application can be determined as follows:

Step 1. Solve for

$$V^1 = \frac{F_b}{(0.036)(SG_H - SG_L)}$$

* Carroll, *Industrial Process Measuring Instruments,* pp. 231–232.

where 0.036 = weight of 1 cu in. of water, lb
V^1 = minimum float volume for a 3- to 15-psi instrument output
SG_H = specific gravity of heavier liquid
SG_L = specific gravity of lighter liquid
F_b = displacement force required to give a 3- to 15-psi output with a proportional-band dial setting, 10

The force is obtained from the displacer shown in Fig. 10-3 and is the style of torque tube to be used.

Step 2. When the minimum volume of the float is known, the required diameter for a given proportional-band setting can then be obtained from the curves shown in Fig. 10-5.

As an example, determine the float size for the following conditions:

1. Length of displacer float desired: 32 in.
2. Specific gravity of the two liquids: 0.98 and 1.02
3. Minimum proportional band desired: 8
4. Style of Level-Trol to be used: top-mounted

$$V^1 = \frac{0.36}{0.036(0.04)} = 250 \text{ cu in.}$$

A 6-in. float is approximately the size required when used with a standard torque tube.

If a light torque tube is used, the required float volume can be reduced by one-half, or

$$V^1 = \frac{0.18}{0.036(0.04)} = 125 \text{ cu in.}$$

which will permit the use of a 4½-in.-diameter float for a proportional band of 8.

Various-size displacer floats are available for interface measurement, but not all are practicable for use under certain conditions.

Differential-pressure-type Element*

PRINCIPLE OF DESIGN: The interface between two immiscible liquids having different specific gravities can be measured very accurately by differential-pressure instruments. The principle is exactly the same as that used for measuring the specific gravity of liquids, the only difference being in biasing the instrument.

Suppose it is desired to measure the interface between two immiscible liquids having specific gravities of 0.80 and 1.30; the measurement is to be made with a diaphragm-type transmitter having a range of 25 in. of water pressure. The required vertical distance between the ends

* *Ibid.*, pp. 232–233.

of the two dip pipes can be found by taking the difference between the two gravities and dividing it into the meter range:

$$1.3 - 0.8 = 0.50$$

$$\frac{25}{0.50} = 50 \text{ in. (vertical distance between ends of dip pipe)}$$

To bias the instrument properly, calculate the differential pressure that would be created if the liquid having the 0.80 gravity covered the ends of both dip pipes, and multiply the vertical distance between them by the specific gravity of the liquid:

$$50 \times 0.80 = 40.00$$

This means that the meter must be biased 40 in. of water pressure for it to read zero when the lighter liquid occupies the space between the ends of the dip pipes. When the heavier liquid occupies this space, the meter would read at the maximum scale marking. A diaphragm-type differential-pressure transmitter such as the one shown in Fig. 10-12 can be biased accurately by connecting an air supply to the high-pressure side of the diaphragm with a 50-in. water manometer in such a way that it will read the pressure on the high-pressure side of the diaphragm. The low-pressure side of the diaphragm must be vented to the atmosphere. The air pressure is adjusted until the manometer reads 40 in. The bias spring is now adjusted until the output-pressure signal reads 3 psi, which is equal to zero differential pressure for a standard 3- to 15-psi transmitter. The meter is now calibrated to measure the interface of two immiscible liquids having specific gravities of 0.8 and 1.30 when the ends of the dip pipes are spaced 50 in. apart vertically.

The available ranges for interface measurement are the same as those for specific-gravity measurement.

CHAPTER ELEVEN

Automatic Control Mechanisms

THE PRECEDING CHAPTERS ON AUTOMATIC CONTROL have been concerned with the theory of automatic control and the control characteristics that result from the use of the different modes of controls available. These modes of control have been described as on and off, throttling, proportional and reset, etc., with little or no mention of the actual mechanical and electrical methods used to produce the control characteristics described. A thorough understanding of automatic control requires not only a knowledge of the types of controls available and the characteristics of the controls resulting from their use on a controlled process but also a knowledge of the actual mechanical details of each type of control mechanism. This knowledge will aid in choosing the right control equipment for a particular process; it will also aid in properly repairing and maintaining the equipment after it is installed.

The study of control mechanisms will be considered as follows:

1. Self-operated control mechanisms
   a. Temperature controls
   b. Pressure regulators
2. Servo-operated control mechanisms

SELF-OPERATED CONTROL MECHANISMS

The simplest forms of control mechanism are those which operate by means of the power available from within the process under control. These controls are simple and cheap and require relatively little maintenance; consequently they are the most widely used. The self-operated temperature controls will be considered first.

Temperature Controls

Self-acting, Pressure-operated Control Valves

The most common type of self-operated temperature control is that shown in Figs. 11-1 and 11-2. In the first illustration the valve stem is directly connected to a bellows filled with a fluid or gas that operates the valve with changes in the temperature of the surrounding atmosphere. In the second illustration the valve stem is operated by a bellows that, in turn, is connected by a capillary tube to a temperature-measuring element that can be placed in a vessel whose temperature is to be controlled.

The valve shown in Fig. 11-1 is used to control atmospheric temperatures and is filled with a liquid having a vapor-pressure temperature curve that causes sufficient expansion of the bellows to close the valve at a predetermined temperature. Although the temperature differential

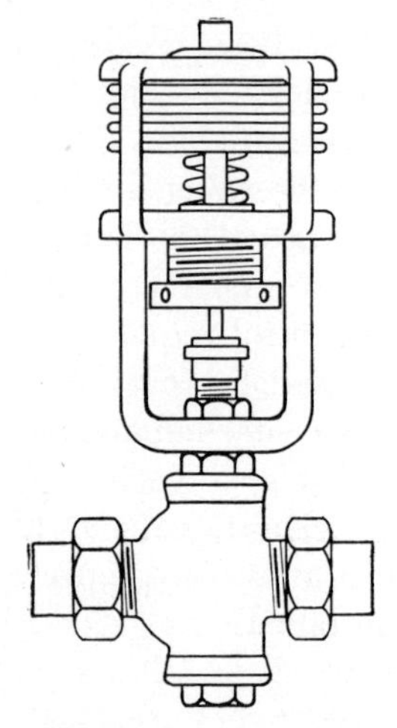

Fig. 11-1 Control valve operated by bellows attached to valve stem.

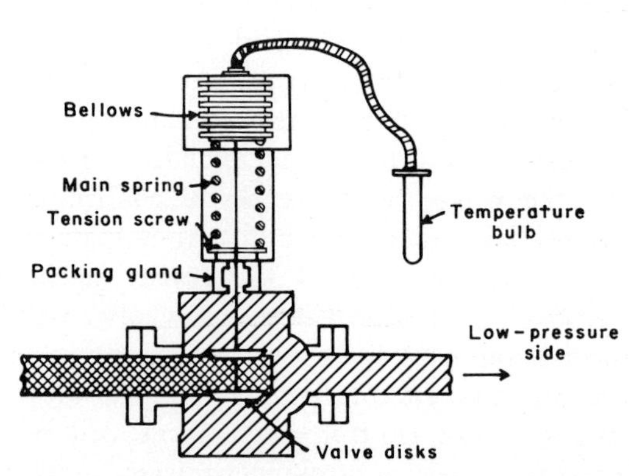

Fig. 11-2 Valve operated by bellows with temperature bulb at a distance.

that must cause a valve of this type to operate is small, it must be of sufficiently large size to produce the power necessary to operate the valve against the steam pressure in the system and the opposing spring.

The control valve shown in Fig. 11-2 has much wider application because its temperature-measuring element can be installed in any process vessel under any temperature condition suitable for a thermometer bulb. The ranges of temperature suitable for this controller are from —40 to +1200°F.

The system may be filled with a liquid (to provide vapor-tension characteristics), a gas, or a solid fluid such as mercury, and it responds exactly as these media respond when used in recording thermometers, which were covered in Chapter 1. Since to be successful a control valve requires an appreciable movement of the valve stem, it is not often possible to use a fluid such as mercury for its operation because the expansion of the fluid is not great enough to operate the valve. Therefore, the control system is filled with either a vapor-tension fluid or a gas, depending upon the temperature. Because considerable power is needed to operate the valve against the pressure in the system, the valve friction, and the valve spring, it is desirable to use a vapor-tension system where temperatures permit. The gas-filled, control-valve system must necessarily be used on special applications where great sensitivity is not needed, since a temperature differential of even 25 to 30°F is often not sufficient to supply the power necessary to operate the valve.

The vapor-tension type of control system is used for temperatures within the range of the available liquids suitable for this use. All the problems of overrange protection, cross-ambient effects, bulb size, bulb location, etc., considered for vapor-tension thermometers, are encountered with this type of control.

Particular use is made in this control system of the limited fill as a means of overrange protection. Since the system is usually considerably larger than a recording-thermometer system, it is easier to provide the exact amount of fluid that produces vapor-tension characteristics over the useful range of the controller and produces superheated gas pressures in the system when heated above the operating range of the controller.

Another characteristic of a controller of this type is the fact that the force available to operate the valve varies with the temperature. Therefore, the valve must operate near the higher temperature range of the vapor-pressure temperature curve of the fluid being used. If it is adjusted by means of the valve spring to operate at too low a temperature, the action will not be positive because of the lack of power available to overcome the friction of the valve stem.

Flash-chamber, Control-valve Operation

The vapor-tension system of control valves is sometimes used in a different way from the conventional vapor-tension systems that were studied in the preceding paragraphs. In this special design the control valve and bellows are always at a higher temperature than the temperature-measuring bulb. This condition would exist where a room temperature was being controlled by means of high-pressure steam. The valve and bellows would approach the temperature of the steam in the line while the temperature-measuring bulb would be at room temperature. The temperature-measuring bulb and the capillary tubing would be filled solidly with fluid, and the bellows or flash chamber would be empty. If the room temperature increases, an expansion of the fluid takes place that forces some of the liquid out of the capillary tube into the flash chamber where it flashes into a vapor and exerts a force to close the valve. If the temperature is reduced, the fluid contracts and makes room for the liquid to return from the flash chamber, which in turn, causes the force exerted in the bellows to be reduced.

This system has the advantage of permitting the use of a small temperature-measuring bulb for large movements of the bellows.

Pressure Regulators

Spring-loaded Pressure Regulators

The self-operated pressure regulator is widely used for providing reasonably constant pressure in chemical processes. The usual type is similar to that shown in the diagram of Fig. 11-3. Here the valve stem is in equilibrium under the two forces acting on the rubber or metal diaphragm connected to the top of the stem. The lower force

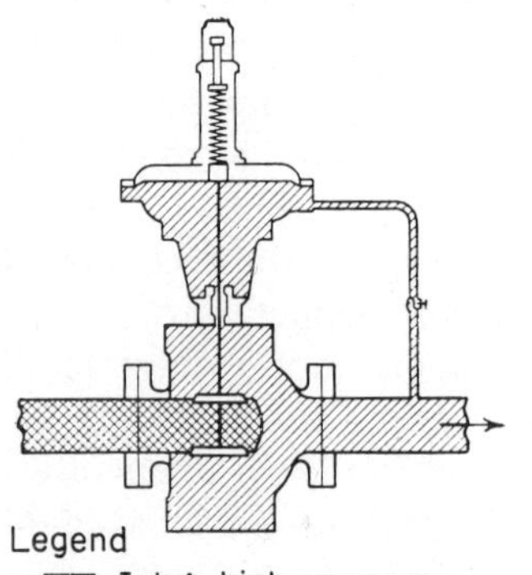

Fig. 11-3 Simple pressure regulator.

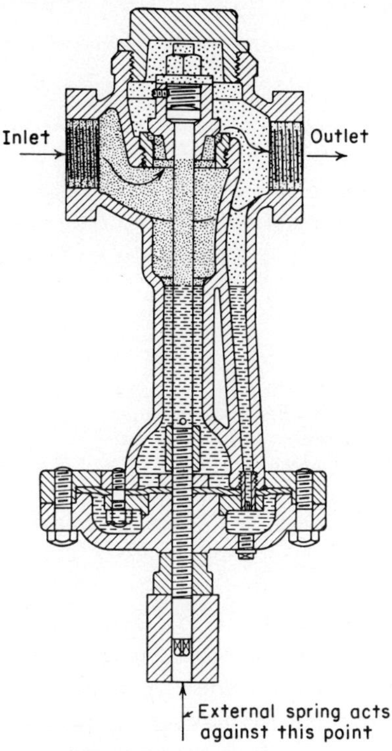

Fig. 11-4 Simple pressure-loaded pressure regulator.

is that of the reduced pressure existing on the downstream side of the valve. The upper force is provided by a spring that can be adjusted to establish the desired downstream pressure. The action of the regulator is obvious. If the downstream-controlled pressure drops below the desired point, the diaphragm becomes unbalanced and moves the valve to an open position. If it is too great, a reverse action takes place, and the valve is closed.

A control valve of this type is essentially throttling in its action; and since there is plenty of power in the system to operate the valve and a minimum lag, it operates without serious hunting, unless the upstream pressures are very unstable.

Pressure-loaded, Self-operated Pressure Regulators

A refinement of the simple pressure-loaded pressure regulator is shown in Fig. 11-4. This regulator is used for high pressures to eliminate the large spring that would be needed to balance the diaphragm under

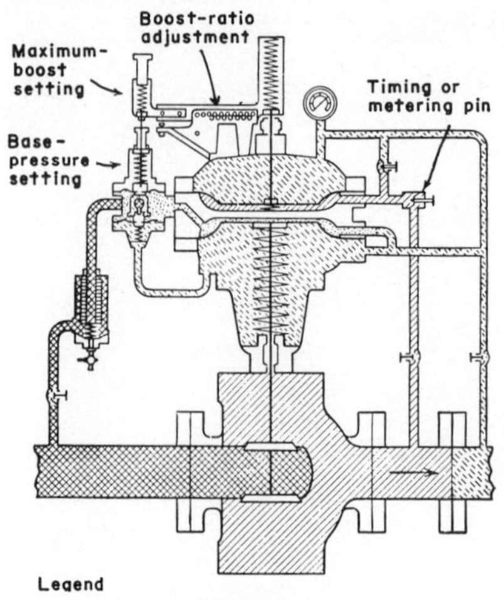

Fig. 11-5 Fisher-King pressure booster. (*Fisher Governor Company*)

this condition. In this regulator the low pressure bears on the top of the diaphragm and the high pressure on the bottom side. A small differential to be balanced by the spring is provided by making the pressure area on the low side smaller than that of the high-pressure side.

Self-operated Pressure Booster

Pressure regulation by boosting the pressure in a gas-distribution system, covered in a previous section, is obtained by changes in the velocity of flow through the line. The Fisher-King gas-pressure booster shown in Fig. 11-5 can be seen to be a modification of the pressure-loaded type. Two diaphragms are provided as well as an orifice on the downstream side which is used to detect deviations in the flow of gas through the system. The operation is essentially as described in the following paragraph.

The system is under a balance provided by the two separate diaphragms, one loaded by the pressure from the low side of the orifice and from the reduced upstream pressure, the other by the upstream and downstream pressures of the orifice. Any change in gas velocity creates on the top diaphragm a pressure differential which in turn varies the adjustment of the pilot. This causes the control valve to move further open or closed to correct for the change in flow. Adjustments are provided for the magnitude of the correction that results from a given change in flow as well as adjustments for the desired amount of reduction in pressure.

SERVO-OPERATED CONTROL MECHANISMS

Servo-operated control mechanisms were developed to overcome the inherent limitations in the self-operated control. Servo-operated, or auxiliary power-operated, instruments are divided into two groups, depending upon the form of energy used to operate the controls. The first and most significant group of controls are operated by compressed air; the second, by electricity. Air-operated controller mechanisms will be studied in this chapter.

Pneumatic Controller Mechanisms

Bailey Meter Company

PRINCIPLE OF DESIGN: Figure 11-6 is a view of the Bailey Meter Company Mini-Line* 500 indicator-controller, which is a combination H/A station and controller. The instrument, designed to receive pneumatic signals, produces the desired control action and necessary indications for the control of complex processes.

The instrument is equipped with the necessary components for automatic and manual control with switching devices and an indicating meter mounted on the front panel. The unit can be equipped with proportional, proportional-plus-integral, proportional-plus-derivative, or proportional-plus-integral-plus-derivative control actions. These adjustments can be installed in any combination. Figure 11-7 is the control mechanism with its pneumatic connections. The controller plugs into the rear of the H/A station, thus eliminating interconnecting piping requirements.

* Registered trade name, Bailey Meter Company.

Fig. 11-6 View of Mini-Line indicator-controller. (*Bailey Meter Company*)

The following will help the student to understand the functions of the various components of the control mechanism shown schematically in Fig. 11-7. The computing (or measuring) mechanism consists of: *A*, *B*, *C*, and *D* bellows; *AB* and *CD* beams; *AB* and *CD* beam springs; a vane-positioning sector; a gain adjustment; an input-reversing switch; a switch to add integral actions; an integral plug-in unit (if desired); and a derivative plug-in unit (if desired). The transmitting portion of the mechanism consists of a booster unit, a nozzle, and a vane.

Pneumatic pressures in *A* and *B* bellows produce forces which position the *AB* beam. The AB beam spring may be in tension or compression to bias the *AB* beam forces. The *CD* beam is positioned in a manner similar to that of the *AB* beam; i.e., through summation of *C* and *D* bellows forces and the *CD* spring force.

Movement of the *AB* beam positions the vane-positioning sector, thus changing the distance between the vane and nozzle since the vane "rides" on the sector. The amount of the vane movement toward or away from the nozzle depends upon the location of the gain adjustment along the vane-positioning sector.

A change in the vane-nozzle relationship results in an increase or decrease in the nozzle back pressure since air escapes through the nozzle at a rate dependent upon the distance between the vane and the nozzle. Normal nozzle back pressure is approximately 2 psig. Small changes in the nozzle back pressure, caused by movement of the vane, are sensed by the booster unit, resulting in an increase or decrease in the booster output pressure, thus increasing or decreasing the output signal from the controller. In addition, booster output pressure is fed back into the *D* bellows, thus repositioning the *CD* beam and thereby repositioning the vane-positioning sector, restoring a normal vane-nozzle relationship.

The control operates on the assumptions that the *B* bellows and the *AB* beam spring act against the force of the *A* bellows. The *C* bellows and the *CD* beam spring act against the force of the *D* bellows. When an equation for each function is to be solved, the base point must be considered. (The base point is defined as: the combination of pressures in the *A*, *B*, *C*, and *D* bellows and the equivalent pressures of the *AB* and *CD* beam springs under initial or calibration conditions so that the gain adjustment can be moved through its entire travel with no change in output from the *D* bellows.) This is the condition around which the controller operates.

The Bristol Company, Division of American Chain & Cable Company, Inc.

PRINCIPLE OF DESIGN: Figure 11-8 is a three-mode advance design series 502 A/D controller. The three modes consist of proportional band

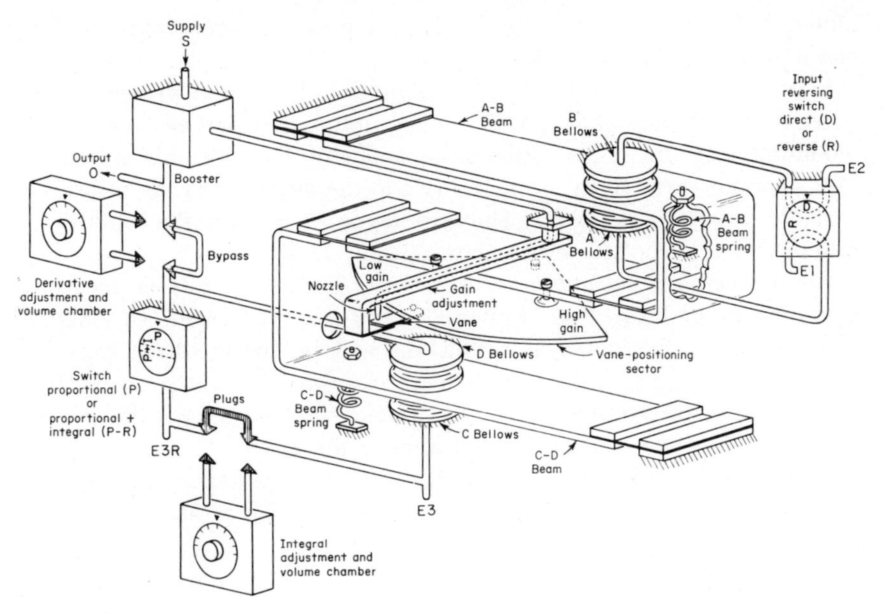

Fig. 11-7 Control mechanism of Mini-Line controller. (*Bailey Meter Company*)

plus reset plus derivative as the controller appears outside the case. Figure 11-9 is a view of dual controllers installed in one of the large Bristol cases.

To understand the operation of the mechanism, refer to Fig. 11-10, which is a proportional control unit of the mechanical-input type. Shown schematically in Fig. 11-11 are some of the major components of the A/D controller with the interconnecting signals between various components. A cutaway view appears in Fig. 11-12.

Measuring element. The measuring element can be any standard,

Fig. 11-8 Three-mode A/D controller. (*The Bristol Company, Division of American Chain & Cable Company*)

high-quality Bristol measuring element for pressure, temperature, differential pressure, flow, or level.

Primary error detector. As shown in Fig. 11-10, the primary error detector is not part of the A/D control unit but is part of the pointer mounting assembly. The differential beam of the assembly moves to give a motion-error signal proportionate to the deviation of the measured variable from the set point. The error signal is fed into the input-lever arm on the A/D control unit, as shown in Fig. 11-10.

Adjustable-gain, secondary-error detector. The input lever moves a *C*-shaped input beam that is pivoted along axis *BB*. Also, a *C*-shaped feedback beam is pivoted along the *AA* axis by the feedback capsular element (see Fig. 11-10). The baffle rocker arm is pivoted on the two beams which are arranged to form a circle. Rotating the baffle rocker arm places its pivot points in any position on this circle.

When the pivot points are near the *BB* axis, the input motion is at a minimum, and the feedback motion to the baffle is at a maximum.

Fig. 11-9 Dual A/D controller. (*The Bristol Company, Division of American Chain & Cable Company*)

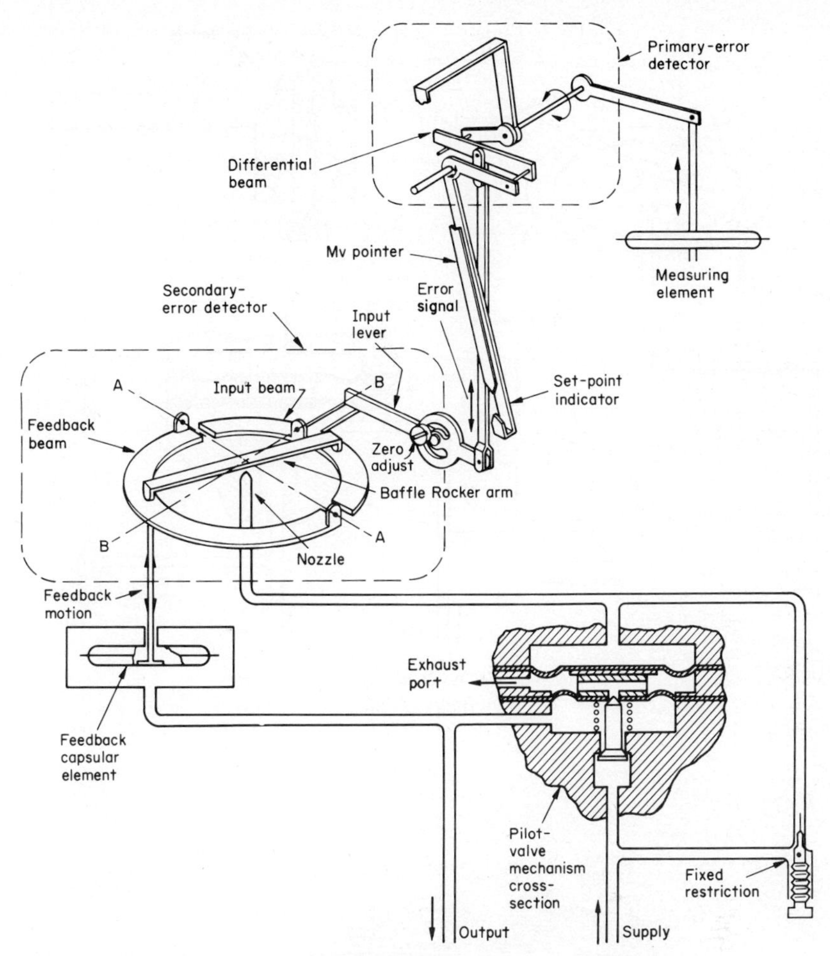

Fig. 11-10 Schematic of mechanical input to A/D controller. (*The Bristol Company, Division of American Chain & Cable Company*)

This provides low gain or a wide proportional band (approximately 400 percent).

With the pivot points near the *AA* axis, input motion is at a maximum, and the feedback motion to the baffle is at a minimum. This provides high gain or a narrow proportional band (approximately 1 percent).

Because the drum-type cover of the A/D control unit is attached to the baffle rocker arm, rotation of the cover rotates the baffle rocker arm and changes the proportional band of the controller. Moving the pivot point across the *BB* axis changes the unit from direct- to reverse-acting.

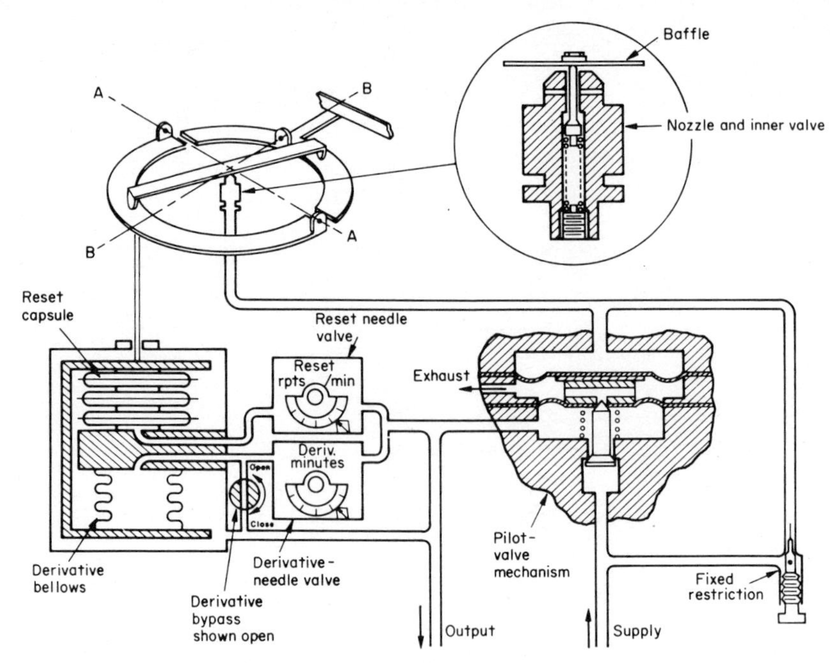

Fig. 11-11 Schematic of three-mode A/D controller. (*The Bristol Company, Division of American Chain & Cable Company*)

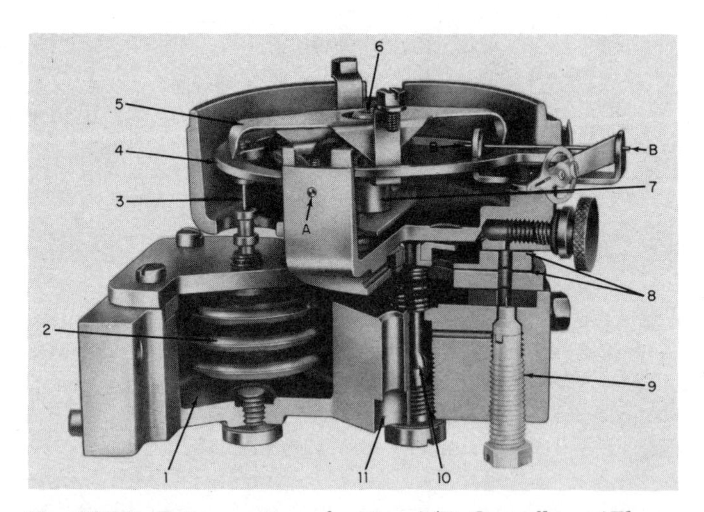

Fig. 11-12 Cutaway view of major A/D Controller. (*The Bristol Company, Division of American Chain & Cable Company*)

PROPORTIONAL-PLUS-RESET CONTROL: A schematic of the proportional-plus-reset unit with internal feedback is shown in Fig. 11-13. Where external feedback is required, the schematic of Fig. 11-14 applies.

The only changes in this unit from the proportional unit are: (1) the nozzle, and the baffle rocker-arm assembly, (2) the feedback-element arrangement, and (3) the addition of the reset needle valve.

Feedback system. The reset-feedback signal is fed to the inside of the feedback element through the reset needle valve. Proportional-feedback pressure connects to the outside of the same element. This combined motion signal from the feedback element goes to the feedback beam of the secondary error detector and reverses the feedback signal from that of a proportional unit. An inner valve assembly is used in the proportional-plus-reset unit.

The reset needle valve, a direct-reading type, is continuously adjustable between 0.02 and 70 repeats per min.

PROPORTIONAL-PLUS-DERIVATIVE CONTROL: See Fig. 11-15 for a schematic diagram of the proportional-plus-derivative control unit. This unit is essentially the same as the proportional-plus-reset unit with two addi-

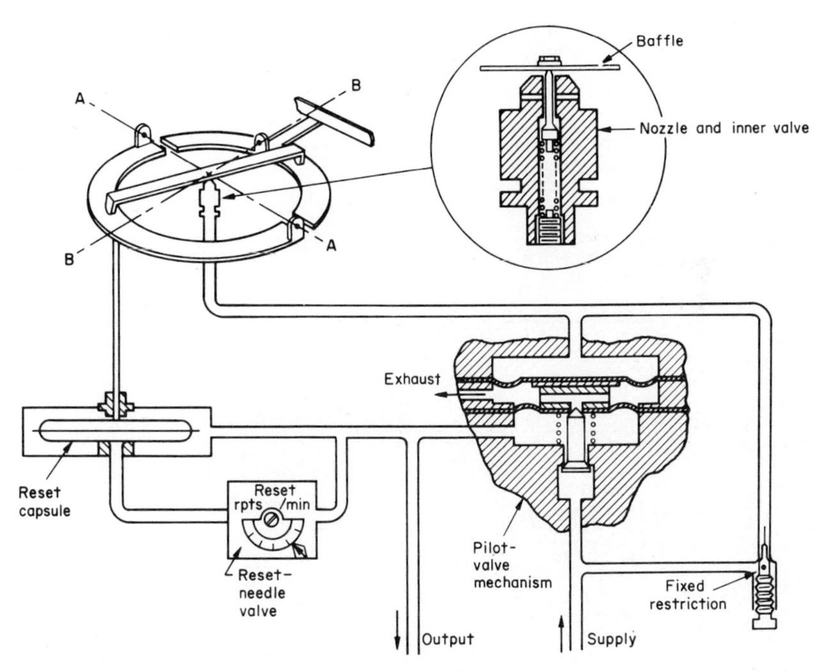

Fig. 11-13 Schematic of proportional plus internal reset, A/D controller. (*The Bristol Company, Division of American Chain & Cable Company*)

tions: a derivative needle valve (range 0.2 to 20 min) and a derivative bellows chamber (see Fig. 11-11).

The derivative-feedback motion is transferred to the proportional-feedback element by a rod. This arrangement and the calibration of the needle valve permits easy adjustment of derivative time settings, which is important to many processes where only small derivative times can be tolerated. Zero-derivative effect can be achieved by opening the needle valve completely.

THREE-MODE CONTROL: Figure 11-11 schematically illustrates the three-mode (proportional-plus-reset-plus-derivative) control unit. In this unit, as in the reset unit, an inner valve with nozzle is used rather than the flapper nozzle of the derivative unit.

The inner valve works in reverse of the flapper nozzle. As the baffle moves downward, it pushes down on the inner-valve pin, and the back pressure decreases. In the case of the nozzle-flapper device, downward movement of the baffle nearer to the nozzle increases the back pressure.

Two feedback elements are used in the three-mode controller and are yoked together so that increasing internal pressures within the elements cause opposing forces. Output pressure from the pilot valve passes through separate needle valves to the two elements (the reset

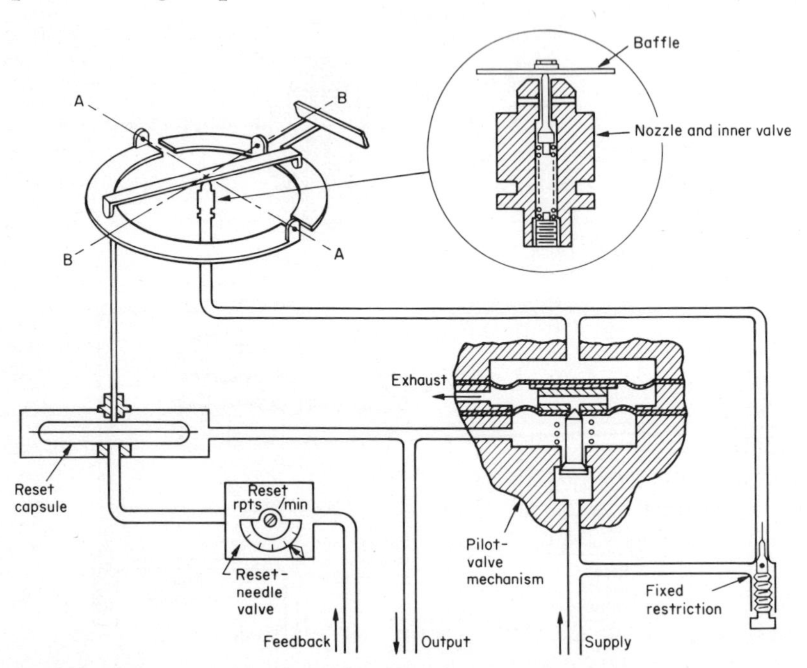

Fig. 11-14 Schematic of proportional plus external reset. (*The Bristol Company, Division of American Chain & Cable Company*)

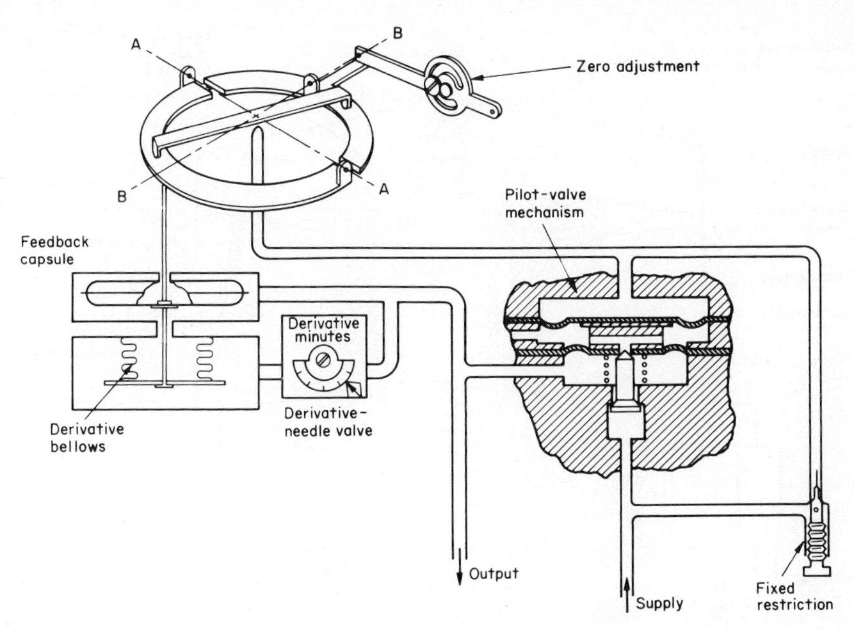

Fig. 11-15 Proportional plus derivative, A/D controller. (*The Bristol Company, Division of American Chain & Cable Company*)

and the derivative) and by a linkage system moves the recording pen. The set point is manually adjusted through a knob-and-linkage system.

In Fig. 11-10, the lower bellows initiates negative feedback response since its action on the feedback pin opposes the initial signal change. The derivative needle valve delays this negative feedback temporarily, thereby producing derivative response.

Action of the upper capsule opposes the lower bellows motion and thereby gives positive feedback. The reset needle valve, by restricting the output pressure fed into the reset capsule, delays the positive feedback, and reset action takes place.

When a process with a three-mode controller is started, it is sometimes desirable to eliminate the derivative action without changing the derivative setting. For this reason, a derivative-bypass valve has been included which can be opened to eliminate derivative action and closed to restore the action.

Fischer & Porter Company*

PRINCIPLE OF DESIGN: Figure 11-16 is a cutaway drawing of a pneumatic controller of the Fischer & Porter Company design.

* Grady C. Carroll, *Industrial Instrument Servicing Handbook,* 1st ed., McGraw-Hill Book Company, New York, 1960, pp. 11-13, 11-14, 11-15.

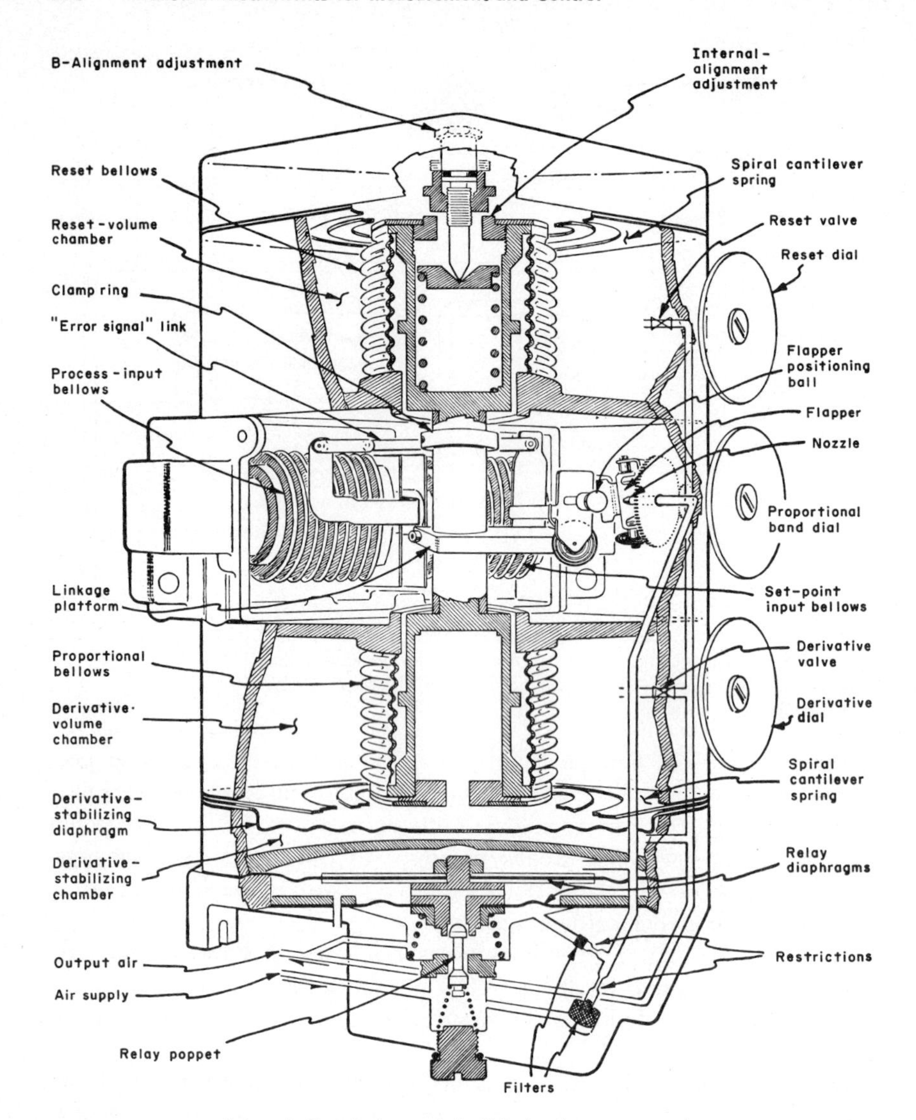

Notes: 1. Except where noted, all text references are made with the controller oriented in the position shown above (vertically).

2. For normal operation the controller is mounted in the horizontal position.

Fig. 11-16 Pneumatic input controller. (*Fischer & Porter Company*)

The unit consists of three major assemblies: a flapper-nozzle pilot, an amplifying relay, and a feedback-bellows system. These assemblies operate to make the controller a motion-balance device.

The input-difference, or error-detection, mechanism is a subtracting device which receives external mechanical signals representing the process variable being controlled and the set point (or "desired value") at which this variable is to be maintained.

The error signal produced by the input-difference mechanism is acted upon by the relay-and-feedback system so that the controller produces an output signal proportional to the difference between the process and set-point input signals.

The pneumatic input controller shown in Fig. 11-16 is used in Fischer & Porter conventional large-case-type instruments where spindle motion representing process and set point can be readily provided. The assembly is designed so that if either the set point or the process is moved with respect to the other, the resulting motion of the output is proportional to the amount of input-link deviation and is in the proper direction. Thus the horizontal position of the flapper ball is determined by the direction and the magnitude of the deviation between the set-point and the process-input signals.

The flapper-nozzle pilot assembly is the motion-to-pneumatic transducing device in the controller. A portion of the air supply passes through a restriction before going to the nozzle. The flapper, positioned over the nozzle, restricts nozzle airflow, thus regulating the nozzle back pressure. For example, if in Fig. 11-16, the flapper-positioning pen (ball) is moved counterclockwise by the error-signal link, the nozzle-flapper clearance is decreased, and the nozzle back pressure is increased.

The proportional gain is achieved by rotating the flapper about the nozzle by means of the proportional-band adjustment; thus, the angular position of the flapper determines the direction and the magnitude of the change in nozzle back pressure resulting from a given error-signal link motion.

The amplifying relay functions to raise variations in the nozzle back pressure to a usable pressure range and to provide the capacity required to operate the final control element. The double-seated valve construction produces a nonbleed-type relay with low steady-state air consumption.

When other than on-off control action is required, the feedback-bellows system is used to produce the net feedback motion with its gain-modifying effect. This system is composed of two opposed bellows connected with a post and two spiral cantilever springs. The connecting post supports the platform on which the flapper-actuating ball is pivoted. The vertical position of the ball is determined by the feedback system.

Proportional control response of the controller is achieved by directing the relay output to the proportional-bellows chamber and preventing any of the relay output from acting upon the reset bellows. Under these conditions, changes in nozzle-flapper clearance are essentially prevented by the proportional-feedback motion, which reduces the overall gain of the controller. The magnitude of this limiting effect depends upon the angular position of the flapper, which is determined by the proportional-band gain adjustment.

Automatic reset is achieved by permitting the relay output to enter the reset-bellows chamber through the reset valve. If, after a change in error signal with responding proportional response, a deviation of set point and process exists, positive feedback action slowly changes the controller output in a direction to eliminate the deviation.

Derivative-controller response is similarly achieved by adding a throttling valve in the relay-output air passage to the proportional-bellows chamber, making the controller response sensitive to the rate of change of the error signal by regulating the speed of response of the negative-feedback system.

Note that in Fig. 11-16 a restriction is connected between the relay output and the nozzle to provide a regenerative-feedback circuit to the flapper-nozzle pilot system in order to improve the proportioning linearity and static performance of the controller. This increases the nozzle airflow at high-relay, output-pressure levels, thereby reducing the required flapper motion to a small value.

Fischer & Porter Company

principle of design: As shown in Fig. 11-17, the Fischer & Porter motion-balance, pneumatic controller receives pneumatic set-point and process information and, acting upon this information, produces a pneumatic output to position a final control element.

In the flapper-nozzle assembly shown in Fig. 11-17, regulated air supply is passed through a removable filter and restriction before emerging at the nozzle. Assume that an input-error signal has moved the positioning ball to decrease the gap between the flapper and nozzle. This reduction in clearance throttles the airflow rate through the nozzle, resulting in an increase in pressure in the nozzle air-supply passage between the restriction and the nozzle. Reversing the input-error signal has the opposite effect of increasing nozzle-flapper clearance and thus decreases the nozzle back pressure. The entire assembly can be rotated about the nozzle which serves as the axis.

Since final control elements are normally operated at pressures of

3 to 15 psi, the normal back-pressure range of 2 to 4 psi is not suitable for normal use. In addition, the restriction limits the flow capacity at much too low a value to be usable for final control element operation. To meet the final control element flow rate and pressure requirements, a high-capacity, nonbleed-type amplifying relay is used. This relay, as shown in Fig. 11-17, has a common air supply with the nozzle. The effective-area ratio between the two diaphragms provides a 6-to-1 amplification of the nozzle back pressure. The high flow-rate capacity of a double-seated poppet valve provides excellent dynamic control response and yet limits steady-state air consumption to a low value typical of the performance of a nonbleed-type unit.

A negative feedback is accomplished by the proportional bellows. This system is composed of two bellows with a connecting post located at right angles to the input-bellows assembly. Attached to the connecting post is the flapper-positioning-ball platform; thus, vertical movement of the post determines the vertical position of the flapper-positioning ball.

Controller-relay output is directed to both the proportional- and reset-

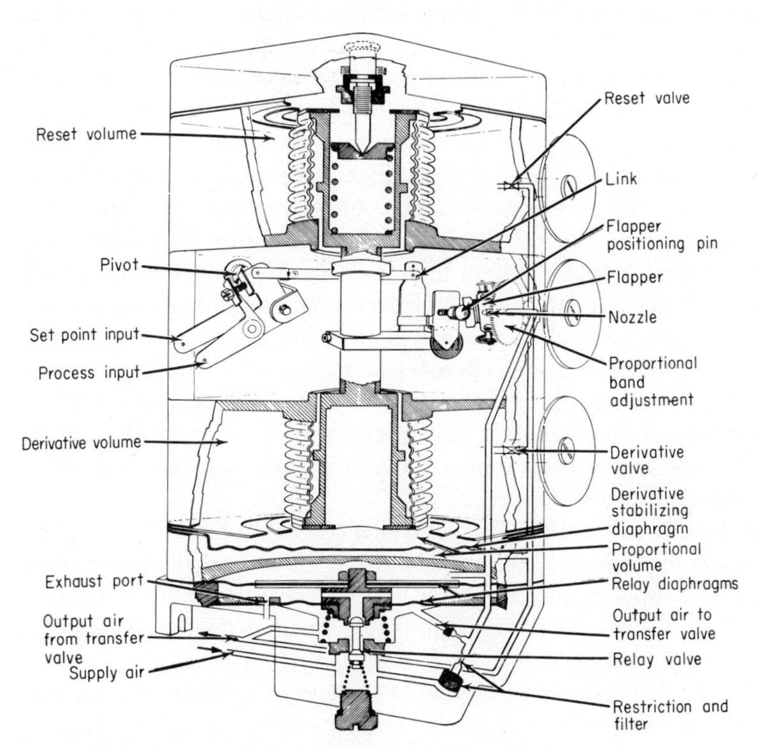

Fig. 11-17 Mechanical input controller. (*Fischer & Porter Company*)

bellows chambers. If there are no restrictions in the relay output passages to both the reset- and proportional-bellows chambers, a change in relay output will not produce any change in the vertical position of the flapper-positioning ball since this new static-pressure condition acts upon both bellows, producing equal and opposite forces but resulting in no movement from the null position. Therefore, no effect from feedback results, and the controller output is unaffected by the feedback system; thus, because of the high-pilot gain, the controller still performs in an on-off manner. If the relay output was transmitted to the proportional-bellows chamber only, the reset-bellows chamber, being either vented to atmosphere or subjected to a constant air-loading pressure, changes in relay output would affect the vertical position of the flapper-positioning ball because of the resulting change in the differential pressure across the feedback system. Therefore, the vertical position of the ball is directly related to the pressure in the proportional-bellows chamber.

To better understand feedback action with the flapper in the position shown and with equal pressures in both the process and set-point bellows of the input-bellows assembly, assume that an increase in pressure is made in the process bellows due to a process change. This increase moves the flapper-positioning ball horizontally to the right. This change in the location of the ball increases the clearance between the flapper and the nozzle, resulting in a decrease in nozzle back pressure. This decrease in nozzle back pressure produces a corresponding decrease in relay output. This decrease in relay output in turn decreases the pressure in the proportional-bellows chamber, resulting in a downward movement of the feedback system and thus the flapper-positioning ball. This downward movement of the ball decreases nozzle-baffle clearance, thus increasing nozzle back pressure. Notice that the negative feedback system (proportional bellows) operates in such a way that it decreases pilot gain. In addition, the combination of input and feedback motions produces a separate and distinct ball position for each controller output value and, in effect, operates in a manner to position the ball along the diagonal path of the flapper.

The reset bellows enables the controller to control at the set point without offset, regardless of controller output. The controller reset-time adjustment is simply a throttle valve in the controller output passage to the reset chamber. Changing this valve opening effects the rate at which reset action takes place.

Derivative action influences controller output by considering the rate at which the process is changing with respect to the set point. Derivative action is accomplished by introducing a restriction in the relay-output passage to the proportional-bellows chamber. In effect, this restriction temporarily eliminates proportional-feedback action upon rapid changes of the set-point-process relationship. This temporary suspension of feedback action produces a corresponding overshoot of controller output due to the high gain of the nozzle-baffle system. As the rate of change of the input air signal decreases, the pressure differential between the proportional-bellows chamber and the controller output passage (differential pressure across the derivative valve) approaches zero, and normal negative-feedback action results.

Fischer & Porter Company

Principle of design: Figure 11-18 is a Fischer & Porter miniature strip-chart recorder-controller which accepts a pneumatic 3- to 15-psig process-variable signal as the input, records it on a 4-in. rectilinear strip chart, and controls this process signal at a predetermined value. A three-position transfer station, including a final control element pressure gauge, permits the selection of automatic or manual operation of the final control element. When specified, a second process-variable signal can be recorded in addition to the controlled process. This instrument is designed for flush-mounting on a panel board. The pneumatic controller furnished with the instrument can be locally mounted (plugged in) on the rear of the case or field-mounted (remote) adjacent to the final control element.

The set-point regulator may be built in the instrument or provisions made for an external set point, thus the recorder-controller can be used in ratio or cascade systems when so specified.

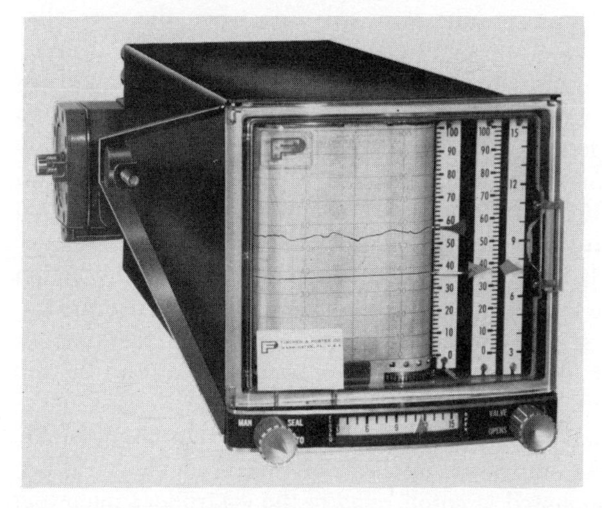

Fig. 11-18 Miniature strip chart recorder-controller. (*Fischer & Porter Company*)

The Fischer & Porter miniature strip-chart recorder-controller consists essentially of: (1) an instrument case, (2) a chassis containing the pneumatic receiving elements and recording mechanism, (3) a three-position transfer station to select automatic or manual operation of the final control element, (4) a built-in gauge to indicate the pressure on the final control element, and (5) a pneumatic controller which may be locally or remotely mounted.

As the process signal enters the capsule, it passes through an adjustable restriction or damping valve. This damping valve permits the process recording to be smoothed to the desired degree. The process signal to the controller originates at the air-connection manifold and is therefore not damped, providing more sensitive control.

The chart-drive motor may be electric or pneumatic. When an electric motor is used, it is protected by an internal 1-amp 250-volt fuse. An optional, mercury, disconnect switch makes the instrument with an electrical chart drive suitable for class I, group D, division II locations. The pneumatic drive is a pulse type requiring a master pulser. One pulser may be used with as many as 25 recorders.

The three-position transfer station is mounted on the inner face of the air-connection manifold. This station consists of a manually adjusted pneumatic pressure regulator and a three-position transfer valve. The two knobs on the face of the recorder-controller extend through the case to permit operation of the transfer valve and regulator. A gasket located between the transfer valve and the air-connection manifold can be rotated to Local or Remote position to correspond to the controller location. A lever, accessible from the rear of the case, is used to change control action (direct or reverse) when the controller is locally mounted.

The transfer valve permits the operator to select the mode of operation desired. The recorder-controller is equipped with a three-position transfer valve which permits selection of Automatic- or Manual-position operation. The intermediate Seal position is used to balance the regulator output with the final control element pressure or the process-signal level so that the transfer of operation does not disturb the process. Rotation of the switch knob actuates the poppet valves by the action of the cam shaft. The transfer valve is directly connected to the air-connection manifold; thus, the transfer-valve position alters the path of air through the manifold to complete the connections required for the mode of operation selected.

The Manual position of the transfer valve permits the output of the regulator to be directly impressed on the final control element. When the controller is field-mounted, a cutoff relay in the controller's mounting manifold isolates the controller output from the final control element.

The Seal position is used to switch the controller from automatic to manual or vice versa. In this position, the controller output signal is isolated from the control system, and the final control element signal is locked in. This permits the operator to view the final control element pressure on the horizontal pressure gauge below the recorder chart. This pressure is used as a reference point when the operator is transferring from automatic to manual or vice versa so that the transfer of operation does not disturb the process. The set-point indicator always indicates required output. When the pressure is being transferred from automatic to manual, it is viewed so that the pressure regulator can be adjusted to the same value as the final control element pressure. When being transferred from manual to automatic, this pressure is viewed so that the pressure regulator, which is now the set-point pressure, can be adjusted to read the exact value indicated by the process pen.

The Automatic position allows the controller output to be connected to the final control element. The pressure-regulator output becomes the set-point pressure and is disconnected from the final control element as it was in the Manual position.

Briefly, the pneumatic controller receives set-point and process information on the input-difference mechanism which operates the pneumatic detector (flapper nozzle). The back pressure of this detector circuit regulates the output pressure of the controller. The controller output signal positions the final control element.

The Foxboro Company

PRINCIPLE OF DESIGN: Figure 11-19 is a front view of the Foxboro series 120 pneumatic Consotrol* recorder which provides a full 4-in.-wide

* Registered trade name, The Foxboro Company.

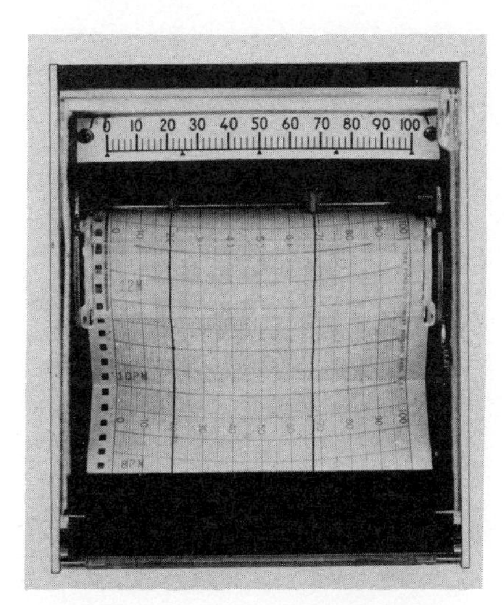

Fig. 11-19 Consotrol recorder. (*The Foxboro Company*)

record equipped with either a roll-type or the unique Foxboro Scan-Fold-type chart drive. The recorder is available in a one-, two-, three-, or four-pen configuration. The input signal to the recorder is the standard 3- to 15-psi value. The recorder can be equipped with signal lights or relay contacts.

The input signal is first passed through an adjustable damping restrictor. The output of this restrictor is the input to the receiver-bellows assembly. The receiver consists of a heavy-duty-impact, extruded-aluminum can containing a large brass bellows working in compression. The large effective area of the bellows assures an extremely linear pressure-to-pen-position relationship. The motion, created by the 3- to 15-psi input signal, is picked up by a conventional link and transferred to an arbor. Calibration is easily accomplished by turning zero, span, and angularity adjustments on this simple linkage system. All calibration adjustments are accessible through a door on the side of the instrument and can be made while the instrument is in operation. The arm interconnecting the pen and the arbor incorporates a rugged tubular member to reduce torsional effects and a takeup spring to reduce mechanical hysteresis. Overrange protection is provided to prevent damage to the bellows or pen-movement assembly for pressures up to 30 psi.

Recorder pens are capillary-fed from easily replaced, plastic ink capsules. The pen in Fig. 11-20 is a precision-molded, high-impact Noryl

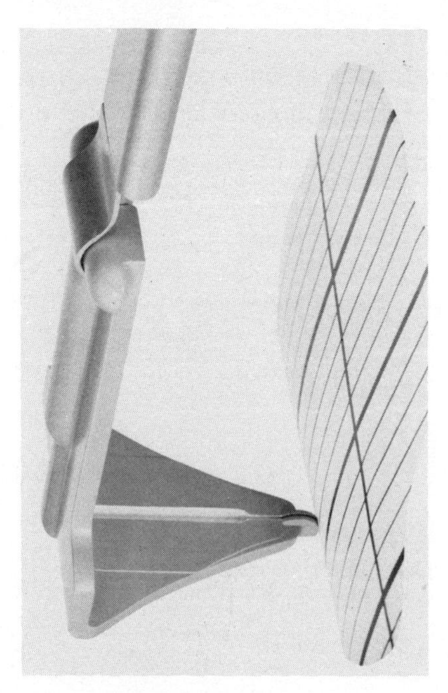

Fig. 11-20 Recording pen for strip charts. (*The Foxboro Company*)

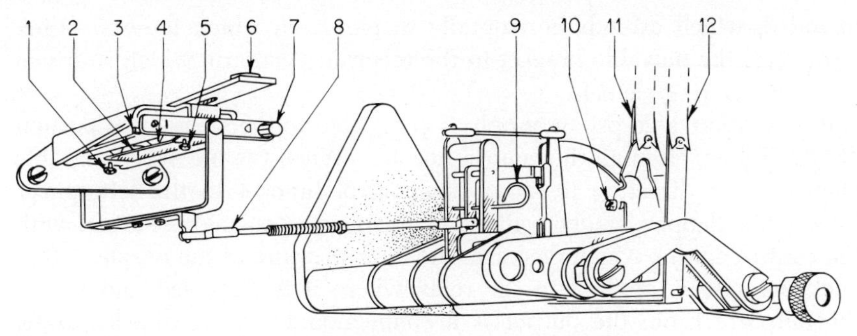

Fig. 11-21 Model 40 narrow-band proportional controller. (*The Foxboro Company*)

plastic design, planned to provide easy-starting, uniform inking on both roll and Scan-Fold charts. The pen arm is made of crushed tubing for torsional rigidity. A pen-tension adjustment is provided on the arbor.

The chassis of the series 120 recorders has been specifically designed for shelf-mounting. Completely enclosed, the chassis consists of a steel box containing an aluminum casting which braces the unit and provides a mounting block for up to four receiver bellows. The internal design provides a "box within a box" to assure rigidity for accurate pen calibration.

A gasketed, removable cover provides easy access to the interior for ink-capsule replacement and adjustment of the damping unit. A door on the side of the recorder provides access to calibration adjustments.

The clear, gasketed, plastic door is designed to reduce glare and improve visibility of the chart and scale.

The Foxboro Company

PRINCIPLE OF DESIGN: Shown schematically in Fig. 11-21 is the control mechanism of a Foxboro Company model 40 narrow-band proportional controller. A similar controller mechanism is used in the Foxboro model 41A narrow-band proportional controller. Only two components not shown in Fig. 11-21 are required to complete the control unit: an automatic manual-transfer switch and an air relay. (They are not shown because they are not important insofar as this discussion is concerned.) The mechanism consists of a control nozzle, feedback nozzle 4, and flapper 2.

Feedback nozzle 4 is rigidly attached to the support bracket. The control nozzle is adjustable in respect to the feedback nozzle by slide 6. Flapper 2 is attached to the flapper bracket. The bracket is pivoted behind adjustment screw 1. The front end of the bracket is attached to rod 3, which extends horizontally to pin 7, on which it rests. Link 8 connects the movable bracket to the recording-pen arm which is driven by the measuring element.

In operation, a 20-psi air supply is connected to the air relay, a portion of which flows through the control nozzle. When the measured variable changes in a direction to move the control bracket to the left, pin 7 lowers the flapper bracket, allowing the flapper to make contact with the control nozzle, which increases the back pressure of the nozzle. This back pressure is applied to the relay where it is amplified and fed to the output. From the output is a connection to the feedback nozzle, which is also located beneath the flapper. The output pressure increases until the flow of air from the feedback nozzle is sufficient to push the flapper away from the control nozzle, at which time the output pressure becomes stable at a different value. Thus, for every position of the recording pen or indicating pointer (depending upon whether the instrument is of the recording or indicating type), there is a corresponding output pressure, the value of which depends upon the setting of proportional band 6.

The Foxboro Company*

PRINCIPLE OF DESIGN: The control mechanism shown schematically in Fig. 11-22 is used in the Foxboro Company's model 40 Stabilog† controller. Not shown in the controller are the proportional and the reset bellows which are not important for alignment of the control components as described below. The controller is designed so that the motion of the recording pen is transmitted to proportional lever 7 and pin 6 to flapper lever 8, which positions flapper 9 in relation to a nozzle. Any movement of flapper 9 changes the back pressure of the nozzle, which is applied to the diaphragm of the air relay where it is amplified and fed to the output from the controller.

A connection from the output feeds the output pressure to the proportional bellows which are beneath and support the bracket on which the proportional lever and flapper are mounted. The reset bellows is mounted above and in contact with the proportional bellows. Proportional-band adjustment 13 rotates flapper lever 8 around pin 6. The setting of the adjustment determines the angle at which flapper lever 8 makes contact with pin 6. With a narrow proportional-band setting, flapper lever 8 is almost vertical. With wide proportional-band setting, flapper lever 8 is near a horizontal position. In a near-vertical position with most of the motion of the recording-pen arm transmitted to proportional lever 7 vertically in response to output-pressure changes, it is

* *Ibid.*, p. 11-17.

† Registered trade name, the Foxboro Company.

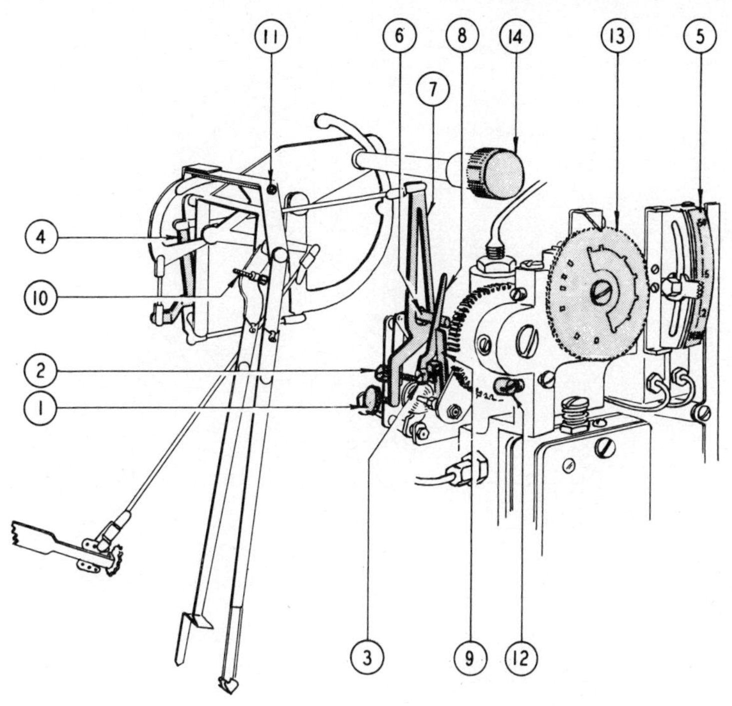

Fig. 11-22 Schematic of pneumatic control. (*The Foxboro Company*)

obvious that large output-pressure changes would be required for the proportional bellows to offset a small change in flapper lever 8, resulting from a change in the measured variable. However, a small change in output pressure would produce a relatively large change in the position of flapper lever 8 if the proportional band was adjusted to position lever 8 in an almost-horizontal position.

The Foxboro Company*

PRINCIPLE OF DESIGN: Shown schematically in Fig. 11-23 is the control mechanism of a Foxboro model 52 indicating receiving controller. The unit consists of a receiving bellows which positions an indicating pointer in response to a 3- to 15-psi pneumatic signal received from a measuring transmitter. (The control mechanism in Fig. 11-26 is attached through a linkage system to the receiving element.) The control mechanism functions to convert the motion of the receiving bellows into a 3- to 15-psi output signal, the value of which depends upon the settings of

* *Ibid.*, p. 11-21.

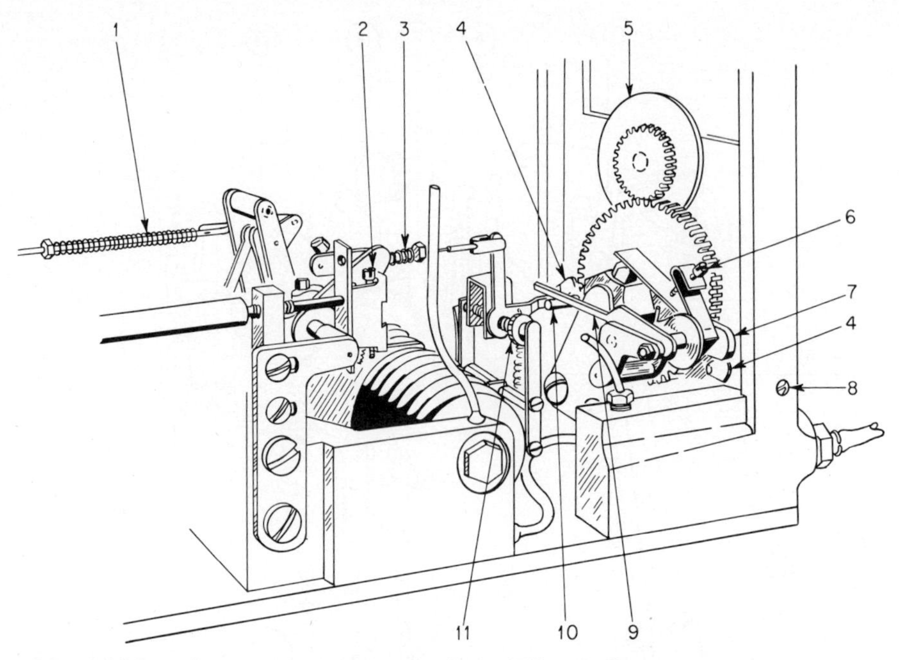

Fig. 11-23 Indicating receiving controller. (*The Foxboro Company*)

the controller and the difference between the set point of the controller and measured variable, as indicated by the receiving-bellows pointer.

In operation, the output of the controller is applied to a controlling device such as a control valve. The controller is constructed in separate units. The air relay and all output connections are attached to the case which is mounted on a panel board. The receiving element and control mechanism is mounted on a chassis which can be withdrawn from the case by removing the Phillips-head screw (located in the center of the front panel below the output-pressure indicator) and the rubber tubes behind the case from the air-terminal block disconnected. The safety latch located at the back of the chasis must be pressed down before the chassis can be completely removed from the case.

The Foxboro Company*

PRINCIPLE OF DESIGN: Figure 11-24 is a drawing of a Foxboro model 59 controller, a force-balance unit which maintains a balance of forces without the use of an amplifying air relay. Two nozzles 3 and 4 are on either side of a baffle. One nozzle vents to the atmosphere; the other supplies air to the bellows housing. The airflow through each

* *Ibid.*, pp. 11-25, 11-26, 11-27.

nozzle is dependent upon the unbalance of forces exerted on nozzle baffle 5 by the bellows assembly. The output pressure of the controller is at the same value as the inside of the bellows housing, which means that the outside of each bellows is exposed to the output pressure at all times.

Beneath the control unit and attached to it is a seal valve which functions to block the air passage from the controller to the control valve when a manual loading pressure is applied to its diaphragm. This provides a means for pneumatically taking a controller out of service from a remotely located manual station so that the control valve can be operated manually from the same station.

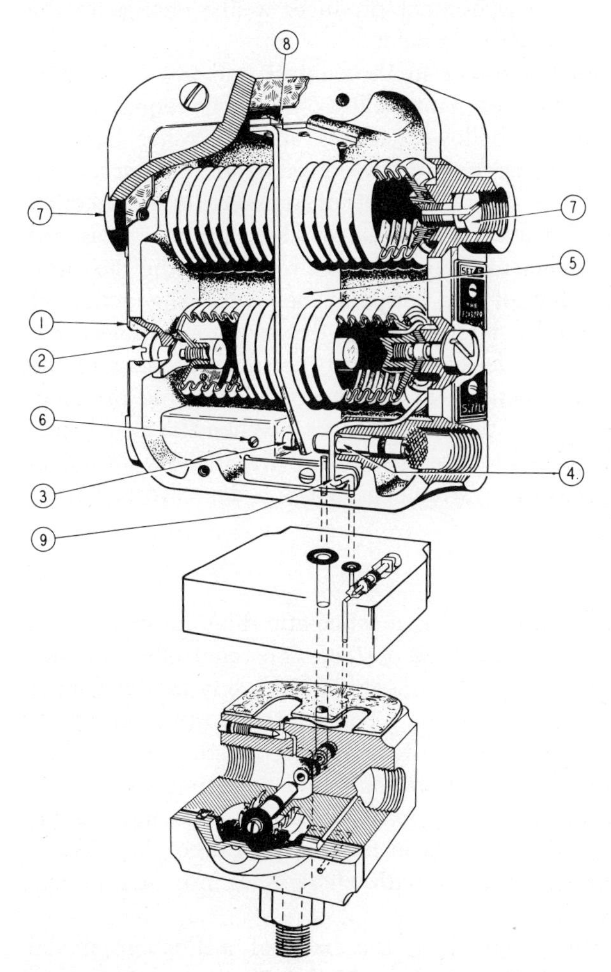

Fig. 11-24 Schematic drawing, control mechanism. (*The Foxboro Company*)

In operation, the output signal from a 3- to 15-psi measuring transmitter is applied to one of ports 7. To the other port 7 is applied a loading pressure which is the set point of the controller. The loading pressure may be from a manual station or from another controller, depending upon whether the controller is a component of a cascade system. The lower bellows on the right-hand side of the nozzle baffle is connected through a needle valve to the controller output and functions as the reset unit of the controller. The bellows opposite the reset bellows is the stabilizing bellows. This bellows has a stabilizing effect on the flapper baffle but does not affect the process. The design of the controller gives it a fixed gain of 1; that is, a change of 1-psi pressure in either the set point or measurement produces a like change in the output pressure when all forces are in balance.

The only time that all the forces in the control unit are in balance is when the pressure in the set-point bellows is exactly equal to the pressure in the measurement bellows. Hence any difference between the set-point pressure and the pressure received from the measuring transmitter produces a change in the output pressure in a direction that operates the control valve to increase or decrease the measured variable by an amount sufficient to produce a pressure in the measurement bellows equal to that of the set-point bellows. The action of the controller is reversed by interchanging the two top connections. When the connections are made according to the black sections of the identifying nameplates, an increase in the measured variable produces an increased output pressure to the control valve. When the air connections are made according to the silver sections of the identifying nameplates, an increase in measured variable produces a decrease in output pressure to the control valve.

The Foxboro Company

PRINCIPLE OF DESIGN: Figure 11-25 is a schematic diagram of a model 58 pneumatic controller of the Foxboro design. The controller operates on the force-balance principle. The "floating disk" acts as the flapper of a conventional flapper-nozzle system. The disk supporting the nozzle is rigid, while the floating disk is free to move around the two fulcrum points which are a part of the proportional-band adjusting lever. The resultant forces due to the upward pressure of the bellows units determines the position of the floating disk in relation to the nozzle. Hence the air-relay output pressure varies with changes in pressure in any of these bellows.

The controller may be mounted on the back of a Foxboro model 53 or 54 miniature strip-chart recorder. Model 54 is a plug-in type where all air connections are made to a receiving manifold; whereas with the 53 model, all connections are made directly to the various components. The control unit may also be installed remotely from the recording unit. As can be seen from the schematic, the controller unit is a pneumatic-set type; that is, the controller's set point is pneumatically positioned by a signal which is manually adjusted at the recorder if the controller is not a part of a cascade or ratio control system. In cascade or ratio control systems, the control point may be set by the output from another controller or a measuring pneumatic transmitter, depending upon the type of system used.

In operation, a pressure is applied to the set-point bellows corresponding to the desired output from the transmitter measuring the process variable to be controlled. The set-point bellows move that side of the floating disk up or down and cause a change in the nozzle pressure, which results in an increased- or decreased-output pressure from the

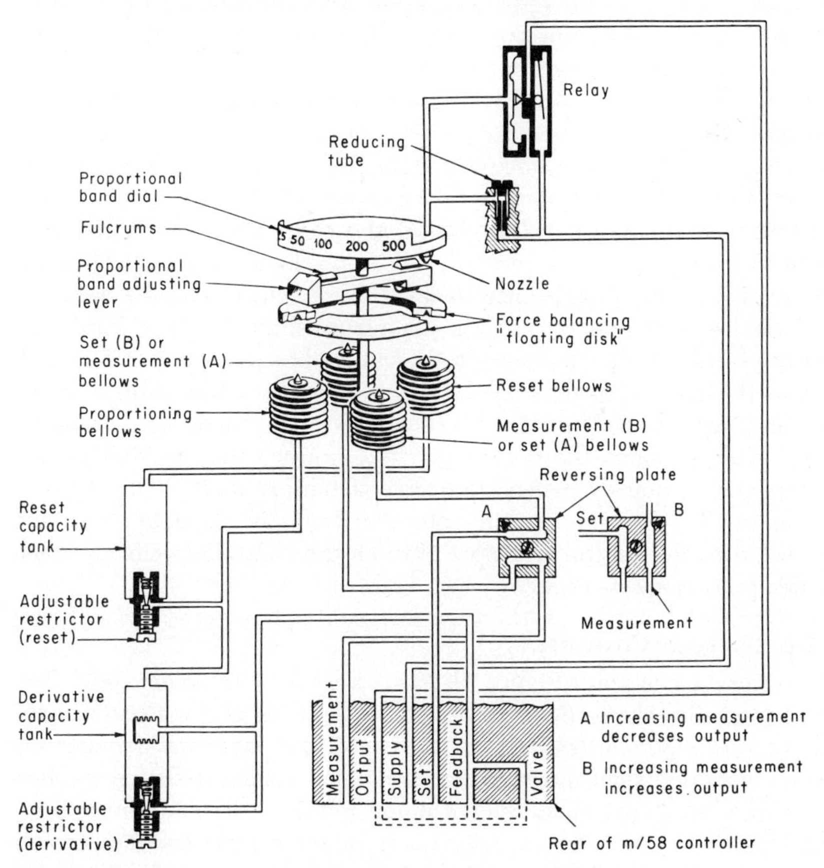

Fig. 11-25 Schematic of a pneumatic controller. (*The Foxboro Company*)

air relay. This variance in output pressure acts to reposition the control valve, thus bringing about a change in the measurement bellows. The variance is also fed back to the proportioning bellows. This continues until a balance of forces is restored against the floating disk. Thus, changes in output pressure are proportional to changes in the set pressure.

The position of the proportional-band adjustment determines the ratio at which the output pressure changes with a change in the set-point pressure. In other words, the gain of the controller is dependent upon the setting of the proportional band. (The gain is equal to 100 divided by the proportional-band setting.)

A reset bellows is also incorporated in the controller. The air circuit is arranged so that the controller's output pressure is fed to the reset bellows as well as to the proportional bellows, but at a rate depending upon the setting of the reset restriction or resistance. This resetting action continues until the pressures in the proportional and the reset bellows are equal. However, this action occurs at an ever-decreasing rate as the final balance point is reached, at which time the measurement and set-point pressures are equal. Thus reset action is dependent upon the deviation of the measurement from the control set point and the setting of the reset restriction.

To further improve the flexibility of the controller, a derivative unit is inserted between the output and the proportional bellows. Thus airflow to or from the proportional bellows, due to changes in the measurement or set-point pressure, causes a pressure drop to occur across the derivative restriction. Pressure in the output line is greater or less than that in the proportional bellows by the amount of this drop. Hence the valve position is determined by the combined proportional-and-reset effect, plus or minus the derivative effect. Any valve position is thus reached more quickly when derivative action is used. The purpose of the small bellows extending into the derivative-capacity tank and connected to the controller output is to ensure controller stability when sudden process upsets occur.

The Foxboro Company

PRINCIPLE OF DESIGN: Figure 11-26 is the series 130 pneumatic Consotrol controller of the Foxboro design. The heart of the Consotrol-100 line of pneumatic instruments is the series 130 pneumatic indicating control station, a shelf-mounted, self-contained station designed to operate with 3- to 15-psi measurement input and 3- to 15-psi output range. The series 130 controller contains many new concepts and design features: (1) the transfer from automatic to manual and from manual to automatic is balanceless and bumpless; (2) the control unit is a high-precision, floating-disk balancing system which provides the most responsive yet stable control action obtainable for fast recovery;(3) there is a full-scale vertical indication of both measurement and set point, together with a unique means of displaying deviation from set point; (4) the automatic section of the controller can be removed from the panel, leaving behind the manual control section to operate the loop or the manual control portion, or the controller can be removed completely from the panel, leaving behind the automatic control portion for operation of the loop; (5) an integral recording mechanism is available as an option to provide a continuous record on a 5-in. circular chart; (6) derivative action is caused by changes in the measurement signal only. There is no derivative effect from a set-point change. A remote-local switch permits the controller to be either pneumatically set by an external 3- to 15-psi source or manually from the front of the instrument. Manual memory pointers are provided on the output indicator for reminding the operator of the normal valve position. The advanced techniques in circuit design and packaging permit improvement of the dependability of the controller.

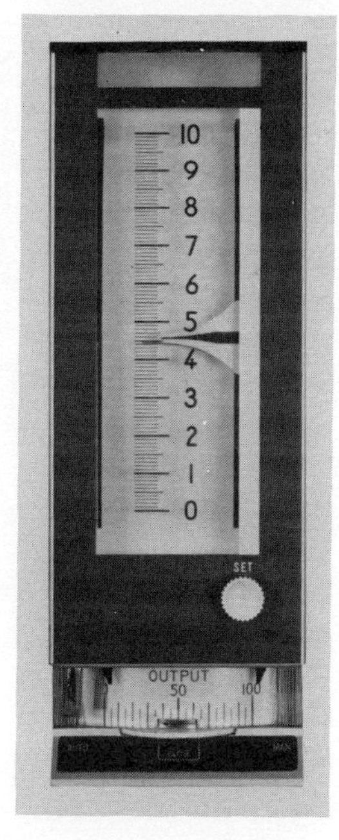

Fig. 11-26 Pneumatic controller. (*The Foxboro Company*)

These features, plus many other options, combined with the versatile shelf-mounting of the entire Consotrol-100 line permit compact, convenient control rooms for the process operator.

The controller is built in two sections. The top section is the measurement and indicating section; the bottom part of the instrument is the controlling part of the device. The design allows either section to be removed without disturbing the other section of the mechanism.

The measurement pointer is a red target ¾ in. high at the right side of the scale, tapering to a point at the scale graduations on the left side of the scale.

The set-point indicator is similar in shape to the measurement pointer but has a white background with a black stripe. When the measurement and set point are within 1 percent of each other, the red measurement pointer is hidden behind the white mask of the set-point indicator. If the deviation is greater than this, a proportionate amount of red appears above or below the set-point indicator to attract the attention of the operator.

Fig. 11-27 Three Consotrol shelf-mounted controllers. (*The Foxboro Company*)

The set-point knob is at the bottom right of the automatic control section of the instrument. Turning the knurled knob mechanically adjusts the manual set point of the instrument while permitting the operator to see the location of the pointer on the vertical scale.

By grasping the bottom of the automatic control unit of the instrument, you can pull the section out of the panel independently of the manual station beneath. Other features are then exposed.

A 5-in. circular-chart recorder, shown in Fig. 11-27, is protected by a clear, plastic door. The signal pen contains its own two-month ink supply. A three-position, chart-drive switch permits the recorder to be turned off or to slow or fast speed. (Fast speed is 60 times faster than the slow speed.)

Pulling the controller all the way out to the stop, as shown in Fig. 11-27, exposes the proportional, reset, and derivative adjustments as well as a reversing switch which allows an increasing measurement to cause either an increasing or decreasing output.

When a release catch on the top of the instrument is pushed, the automatic control section can be completely removed from the panel. Releasing a quick-disconnect plug automatically seals the connections and permits complete removal of the instrument. If the recorder option is used, it is also necessary to unplug a power cord.

On the right side of the automatic control section housing are zero, span, and measurement-damping adjustments. Figure 11-26 is a test switch, identified by the arrow, for easy calibration of the indicating pointers.

The derivative unit. The derivative function in the series 130 controller is located in the measurement signal only so that there is no derivative response to the set-point change. Since derivative action (rate) is applied to the measurement signal, the control unit quickly responds to measurement changes on batch processes without the inherent delays common to units having derivative in the controller's feedback circuit. During steady-state conditions, the unit acts as a 1:1 repeater, but upon a change in measurement, the unit adds a derivative influence to the measurement change.

In the derivative unit, shown schematically in Fig. 11-28, the force moment (bellows area times distance from fulcrum) of bellows *A* is 16 times that of bellows *B*, and the force moment of bellows *B* plus bellows *C* equals that of bellows *A*. As the measurement signal increases, the immediate change in feedback pressure in bellows *B* is 16 times the change in pressure in bellows *A*. Simultaneously, air starts to flow through the restrictor to bellows *C*, gradually reducing the pressure needed in bellows *B* to restore equilibrium. Thus, the output of the derivative unit, which is the signal to the automatic control unit measurement bellows, reflects the change in measurement plus a derivative response added to that change. The graph at the bottom of Fig. 11-28 shows the signal that the measurement bellows in the automatic control unit receives from the derivative unit at the measurement change shown.

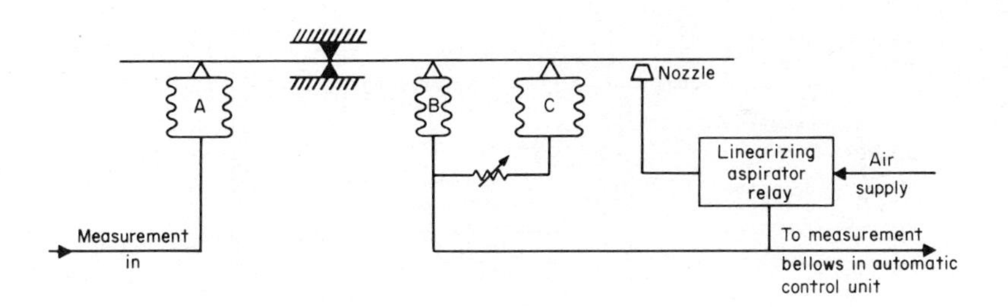

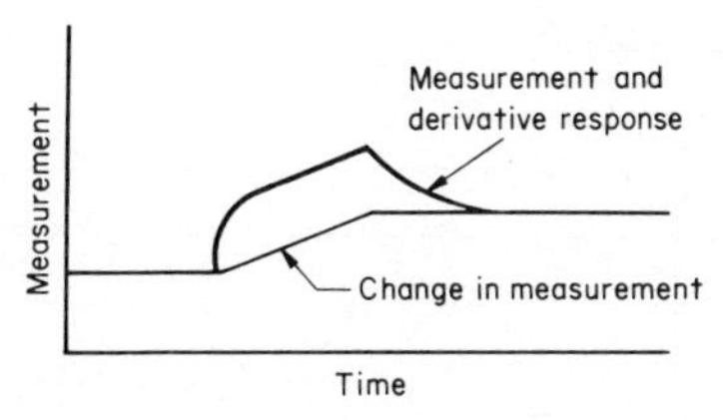

Fig. 11-28 Schematic of Consotrol derivative unit. (*The Foxboro Company*)

Automatic control unit. The automatic control unit operates on the classic force-balance principle. Four bellows—one each for set-point signal, measurement signal, proportional feedback, and reset feedback—act on a floating disk, two on each side of a fulcrum. The resulting summation of forces times moment arm equals zero and keeps the floating disk in equilibrium. The floating disk acts as a flapper of a conventional flapper-nozzle detector system as illustrated in Fig. 11-29.

If the adjustable fulcrum could be positioned directly over the proportional-feedback and the reset bellows, as shown in Fig. 11-29, the unit would have on-off control action. The slightest measurement increase above "set" would cover the nozzle and increase the output to 100 percent. Any decrease in the measurement below set would uncover the nozzle and cause a zero output.

If the adjustable fulcrum is moved to the position shown in Fig. 11-30*a*, throttling control action would result. If distance *B* is four times distance *A*, the unit has a 25 percent proportional band (a gain of 4). If the measurement increases, the output of the relay must increase four times the change in measurement pressure in order to restore

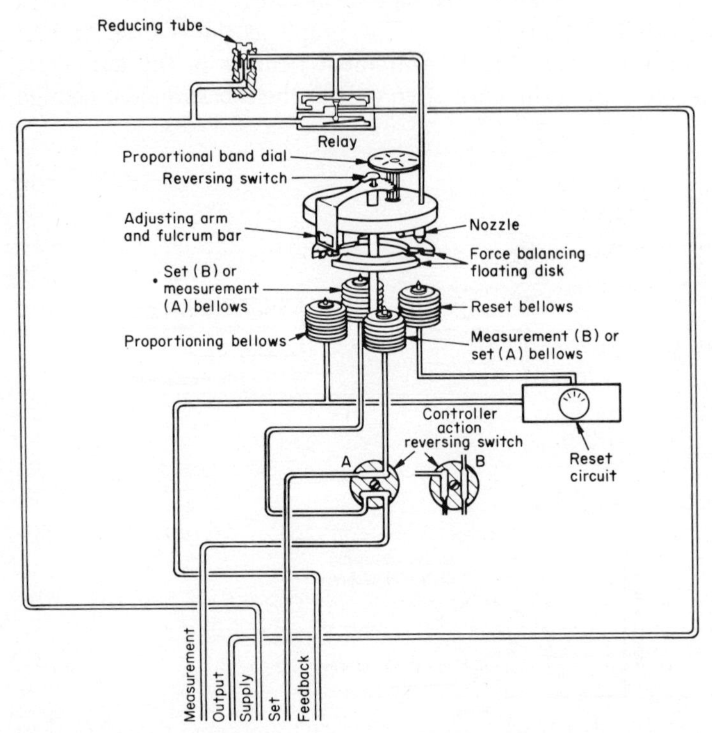

Fig. 11-29 Schematic of Consotrol Controlling mechanism. (*The Foxboro Company*)

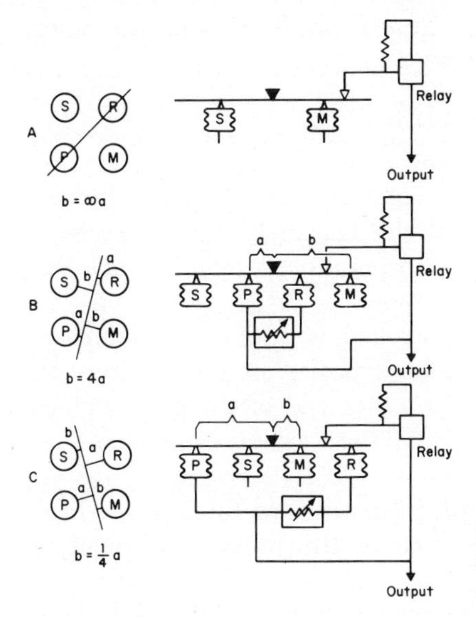

Fig. 11-30 *a, b, c.* Schematic of the adjustable fulcrum. (*The Foxboro Company*)

the balance of forces (if for the moment the reset action is disregarded). If the fulcrum is moved to the position shown in Fig. 11-30*b*, a 400 percent proportional band exists (a gain of ¼) if distance *A* is four times distance *B*. On a change in measurement pressure, the output of the relay must increase only one-forth the change in measurement pressure to restore the balance of forces, again with reset action ignored. The series 130 has a proportional band continuously adjustable from 5 to 500 percent (gain of 20 to 0.5).

Refer again to Fig. 11-29 for reset action. As the measurement pressure increases, the relay pressure and the proportional-bellows pressure must increase four times the change in measurement pressure. However, the output also bleeds through a variable restrictor to the reset bellows, acting on the same side of the fulcrum as the measurement bellows. As the pressure in the reset bellows rises, the output (the pressure in the proportional bellows) must rise even higher to restore the system to balance. A difference in pressure exists between the proportional and the reset bellows until such time as the process reacts to the controller output and lowers the measurement-bellows pressure. This, in turn, lowers the pressure in the proportional bellows until eventually there is no differential between the proportional and the reset bellows. Reset action continues as long as there is a difference between measurement and set, and hence a difference between the proportional and the reset bellows.

Among the other flexibilities of the Series 100 instruments are self-contained alarms, matching the four-pen recorder with various control units, grouping controls, loading stations, ratio controllers, and the auto-selector control system.

Hagan/Computer Systems, Division of Westinghouse Electric Corporation

PRINCIPLE OF DESIGN: Figure 11-31 is a cutaway view of a nonindicating and nonrecording pressure controller of the Hagan/Computer Systems, a Division of Westinghouse Electric Corporation, design. Note that the controller is a proportional-plus-reset type.

In operation, the pressure to be controlled is applied to a bellows pressure element. One end of the bellows post rests on the free end of the pressure-measuring bellows. The other end of the post contacts a beam to which the combined forces of the various components are applied.

The set-point screw is adjusted until the downward force of the loading spring is equal to the upward force of the pressure element on the beam.

An increase in pressure above the set point moves the right-hand end of the beam upward, thereby closing the exhaust port and opening the inlet port of the pilot valve. This action produces an increase in the output pressure to the control device, such as a control valve or piston.

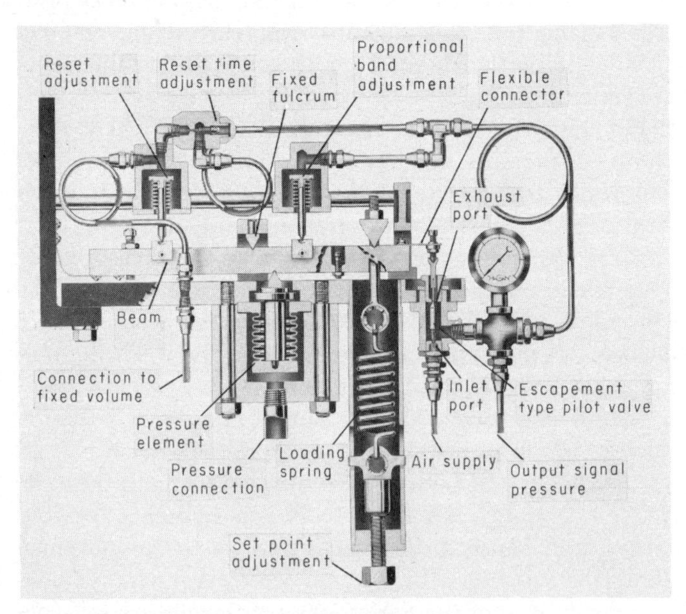

Fig. 11-31 Control mechanism. (Hagan/Computer Systems, Division of *Westinghouse Electric Corporation*)

Through a feedback tube, the output pressure is applied to the proportional-band bellows, whose force opposes that of the pressure element, thus giving rise to the proportional mode of this proportional-plus-reset controller.

Reset action is provided by bleeding a small flow of air from the output through an adjustable restriction into the reset bellows. The force of the reset bellows is applied to the beam in such a way that it raises the pilot valve, which produces the same results as the original increase in pressure on the pressure element.

The proportional and the reset bellows continue to operate on the beam until the controlled pressure applied to the pressure element is restored to the value of the set point or the maximum supply pressure is applied to the control element.

The effective forces of the proportional and the reset bellows are adjustable by moving them with respect to the fixed fulcrum.

Honeywell, Incorporated

PRINCIPLE OF DESIGN: Figure 11-32 is a cross-sectional schematic of a Honeywell, Incorporated, Air-O-Line* pneumatic controller. This instrument provides corrective action based on the *size* of a deviation of the controlled variable from a selected set point (proportional action) and the *time* a given deviation lasts (automatic reset). Reset action automatically corrects for offset (or droop) produced by load or set-point changes. The Air-O-Line controller maintains accurate control over any process having either small or large capacity and demand changes.

Operation of the controller is shown in Fig. 11-32. Air passing through the restriction in the supply line goes to the nozzle of the control unit and to the pilot valve. Back pressure from the nozzle to the pilot valve is regulated by the flapper in the control unit, which moves when the pen responds to a change in the controlled variable. The pilot valve converts small changes in nozzle pressure to significant changes in controlled air pressure. The amplified nozzle pressure is also fed back to a follow-up bellows of the control unit, which repositions the flapper an amount proportional to the deviation of the variable. The second (reset) bellows of the control unit further repositions the flapper, producing a controlled air pressure that compensates for offset due, for example, to load changes.

Instrument control mechanism and differential linkage. Any movement of the pen is transmitted through the differential linkage to the flapper-actuating lever. Thus, any change in the controlled variable

* Registered trade name, Honeywell, Incorporated.

is converted into rotation of the flapper-actuating lever. Rotation of the flapper-actuating lever positions the flapper with respect to the nozzle. The resulting change in nozzle back pressure, amplified by the pilot valve as described below, is immediately transmitted to the follow-up bellows and, through its liquid fill, to the inner bellows. The change in pressure within these bellows horizontally moves a rod connected to them. Movement of the rod, transmitted through an adjustable lever system, repositions the control-unit flapper in the opposite direction an amount determined by the proportional-band setting (see below). This movement of the flapper is just enough to stabilize the controlled air pressure from the pilot valve at the new value required by a change in the controlled variable. The change in controlled air pressure is proportional to the deviation of the pen from the set point.

The proportioning action described above produces a definite position of the final control element for every deviation of the pen from the

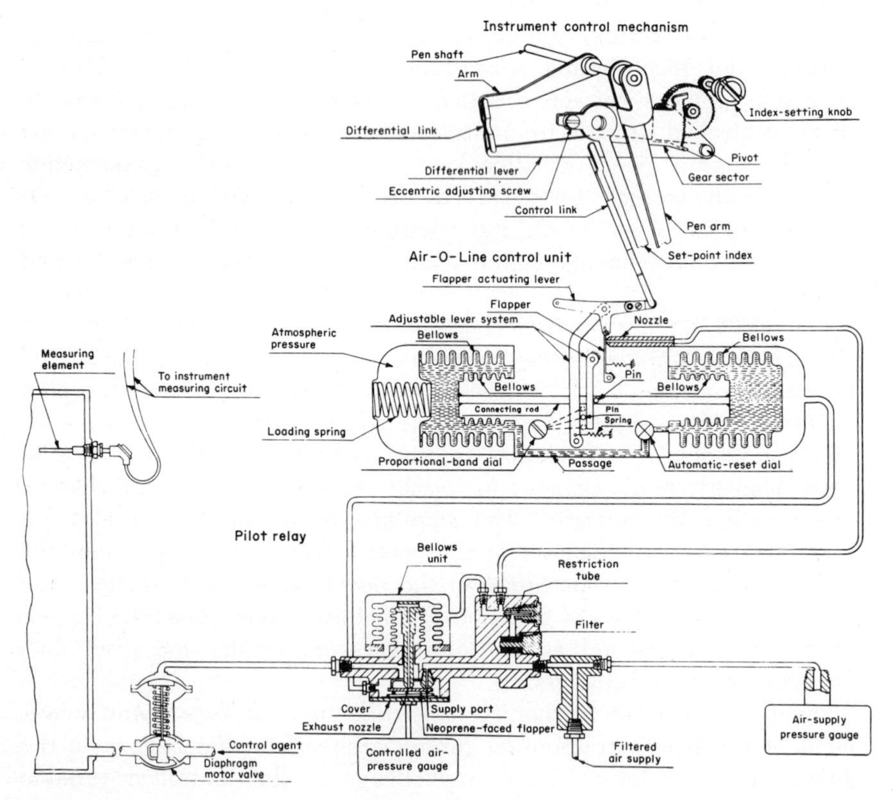

Fig. 11-32 Cross-sectional schematic view of Air-O-Line pneumatic controller. (*Honeywell, Incorporated*)

set point. Without automatic reset, a sustained load change would cause the pen to line away from the set-point index.

Automatic reset repositions the flapper to correct for offset as follows. The liquid fills between the large and small follow-up bellows are connected by a passage to the space between the corresponding large and small reset bellows. Both small inner bellows are spring-loaded and will return to their normal position only when the liquid pressures in the two bellows assemblies are equalized.

As long as offset continues, the pressures in these bellows are unequal, and liquid flows from the follow-up to the reset bellows until balance is restored. The rate of this flow is determined by the setting of the adjustable needle valve (reset adjustment) in the passage between the bellows assemblies. As these pressures are equalizing, the connecting rod is moving back toward its normal position, moving the flapper back in the direction of the initial change.

The net result is that the pen is returned to the index and the controlled air pressure from the pilot valve is increased or decreased, which repositions the final control element to correct for both the size of the deviation of the controlled variable and the length of time it continues.

Changes in nozzle back pressure are applied to the larger (outer) bellows of the pilot valve. Controlled air pressure is applied to the smaller (inner) bellows. When the forces on these bellows balance each other, both the exhaust and supply ports are covered by the pilot flapper.

An increasing back pressure moves the larger bellows downward, forcing the smaller bellows to move with it. The smaller bellows carries the exhaust nozzle, which pushes against the pilot flapper, thus keeping the exhaust port closed and opening the supply port.

A decreasing back pressure relieves some of the force on the large bellows. The reverse movement of both the larger and the smaller bellows opens the exhaust port and permits the flapper to close the supply port. When the forces on both bellows are equalized, the exhaust and the supply ports are again covered by the pilot flapper.

The proportional band (throttling range) is defined as the percentage of full-scale pen movement required to move the final control element from one limit to the other. In Air-O-Line controllers, the proportional band corresponds to a pressure range of 3 to 15 psi. When a reverse-acting controller is used (for example, with an air-to-open control valve), the valve is fully open when the controlled air pressure from the instrument is 15 psi and fully closed when the controlled air pressure is 3 psi.

With narrow proportional bands, a relatively small deviation of the pen from the set-point index represents a relatively large percentage of the band width and thus causes a correspondingly large movement of the control valve. With wide proportional bands, since a large deviation may be only a small percentage of the band width, the movement of the valve is correspondingly smaller.

In an Air-O-Line controller, this relationship is maintained as discussed below. Within the proportional band, each deviation produces a specific change in nozzle back pressure and thus in the feedback pressure to the follow-up bellows. For each pressure change, the follow-up bellows always moves the connecting rod the same horizontal distance. The proportional-band adjustment varies the vertical position of the fulcrum about which the control-unit flapper operates. This varies the distance the flapper is repositioned with respect to the bellows' rod travel and thus for each pressure change. With narrow proportional-band settings, the fulcrum is so positioned that the flapper is moved back only slightly. With wide-band settings, the position of the fulcrum causes the flapper to move back almost to its original position. Consequently, the net movement of the flapper, and the corresponding change in controlled air pressure, increase as the proportional-band setting is decreased. With narrow-band settings, large, frequent deviations of the pen from the set-point index produce essentially on-off control. With a proportional-band setting of 100 percent, and percentage of scale deviation of the pen produces an equivalent percentage of change in control-valve movement. That is, if the pen or pointer deviates 5 percent of full scale, the valve moves through 5 percent of its total travel. If the band setting is greater than 100 percent, the control valve is never fully opened or fully closed. There are certain valve positions beyond the full-scale maximum and minimum pen positions.

With a given proportional-band setting, Air-O-Line control units provide a definite relationship between pen position and controlled air pressure. For example, with a proportional band of 100 percent, if a valve opening of 50 percent is required to maintain the controlled variable at the set point, the controlled air pressure is at 50 percent (9 psi) and the pen is aligned with the set-point index. This relationship holds true, however, only for one set of process conditions. If a sustained-load change occurs or if the set point is changed, the pen lines away from the set-point index to produce the controlled air pressure required to reposition the valve. For example, in controlling the temperature of furnace contents, if the normal charge is increased from two to four tons, a larger valve opening is required to maintain the temperature at the desired set point with the added load.

Reset automatically reestablishes the control level at the desired index setting by shifting the pen-to-controlled air-pressure relationship up or down the scale so that pen position corresponds to the required new pressure at the same index setting.

The Honeywell Air-O-Line controller has been proved over many years of field operation and is very simple in its design and construction. Troubleshooting, maintenance, and testing have been simplified in every way.

Honeywell, Incorporated

PRINCIPLE OF DESIGN: Figure 11-33 is a basic PneumatiK Tel-O-Set* control system of the Honeywell, Incorporated, design. The system consists of three units: a transmitter, a receiving control station, and a controller. The transmitter measures the process variable and sends a proportionate air pressure to the receiver, and the controller controls the final control element.

The receiving control station records or indicates the process variable, transmits a set-point pressure to the controller mounted on the rear of the receiving recorder or indicator, and incorporates provisions for

* Registered trade name, Honeywell, Incorporated.

Fig. 11-33 Pneumatic Tel-O-Set. (*Honeywell, Incorporated*)

manual-automatic switching and manual regulation of air-to-valve pressure.

The controller compares the process-variable with the set-point signal and positions the final control element to maintain the process variable at the desired set point.

When two controllers are interlocked, a cascade control system is made available. These controllers are known as primary and secondary controllers. The output from the primary controller adjusts the set point of the secondary controller, and its output in turn positions the final control element. The use of this system decreases deviations and increases the speed of line out of the process variable after disturbances.

Capsular-control elements are unstressed when the system is in balance and stressed only when there is a change in process variable or set point that initiates control action. This increases the life of the elements and prevents control drift. Capsules are constructed from constant-modulus nickel alloy to minimize changes in accuracy with variations in ambient temperature.

Manual positioning is easily done by first turning the switch lever to the Bal position and then manually aligning the set-point index with the valve pressure-indicating pointer and switching to the Man position. For automatic switching, go to the Bal position and then align the set-point indicator with the process-variable pen or pointer and switch to the Auto position. It is not necessary to match the guage pressures; this is done when the set-point index is aligned. The precise alignment is accomplished with a micrometer set-point-index adjustment.

The set-point index can be positioned either with or without the micrometer mechanism. When the index is depressed, it can be freely moved to any position along the scale. When released, a micrometer mechanism is engaged, and precise setting is made by turning the micrometer adjusting the thumb wheel at the right-hand end of the scale.

The cases and chassis of the single pointer and the single record-control stations are interchangeable. Air connections are the quick-connect type; no tubing connections need be made. An indicating chassis can be operated in a recorder case; a recorder can be operated in an indicator case, but extra provisions must be made for power to the chart being driven.

A separate unit is the control station containing the set-point and valve-pressure mechanisms, and indication permits complete removal of the recording or indicating chassis while on either automatic or manual control.

Pen-zero and -damping adjustments are accessible from the front of the unit without withdrawing the chassis. The effects of the damping adjustment on the chart can be seen while adjustments are being made. Zero adjustment on the indicating control station is located on the pointer and can be made without removing the chassis from the case. Indicator Bourdon gauges have fixed damping.

Calibrated adjustments at the rear of the control unit enable precise setting of proportional-band, reset, and rate values. Front-set control adjustments are available as an option and are accessible with the door open. The back-set adjustments remain on the controllers supplied with front-set option.

The proportional band is adjusted over a double scale, graduated in both percentage-proportional band and gain.

Either fast or slow ranges of reset and rate can be obtained by a screw adjustment. The rate unit can be completely cut out of the circuit by loosening three screws and rotating the rate unit.

In a single-station cascade system, the functions of the primary and secondary control stations are combined in a single two-pen recorder or a two-pointer indicator. The instrument records or indicates both the primary and secondary variables and uses the output of the primary controller to adjust the set point of the secondary controller. The primary controller is integrally mounted with the control station but can also be separately mounted if desired. The secondary controller is separately mounted.

There are some applications where it is desirable to use a two-station cascade control system. For example, an application where the secondary control station must be at a different location on the panel for convenience of operation, or where another record or indication is needed in addition to those of the primary and secondary variables.

In a two-station cascade system, the primary variable transmitter sends its signal to a standard recording or indicating control station. The primary variable controller can either be integrally mounted with the control station or separately mounted. The primary control signal adjusts the set point of the secondary control station. This station records or indicates the secondary process variable and controls it in accordance with changes in the output of the primary controller.

Moore Products Co.*

PRINCIPLE OF DESIGN: Figure 11-34 is a model 55 Moore Products Co. Nullmatic† stack controller with an integral-control, set-point regulator. The controller operates on the pneumatic null-balance principle to control a process variable in response to a pneumatic signal received from

* Carroll, *Industrial Instrument Servicing Handbook*, pp. 11-64, 11-65, 11-66.

† Registered trade name, Moore Products Co.

a measuring transmitter. The entire controller works on a balance of forces without the use of levers and pivots.

The five major components of the unit are set-point regulator 1, control-diaphragm assembly 2, air relay 3, reset adjustment 4, and proportional-band adjustment 5. Within the body casting is an air passage which permits the control set-point pressure from the set-point regulator to be applied to either the top or bottom section of the control-diaphragm stack. The pneumatic signal from the measuring transmitter can also be applied to either end of the stack. The signals from the set-point regulator and the transmitter are always connected to opposite

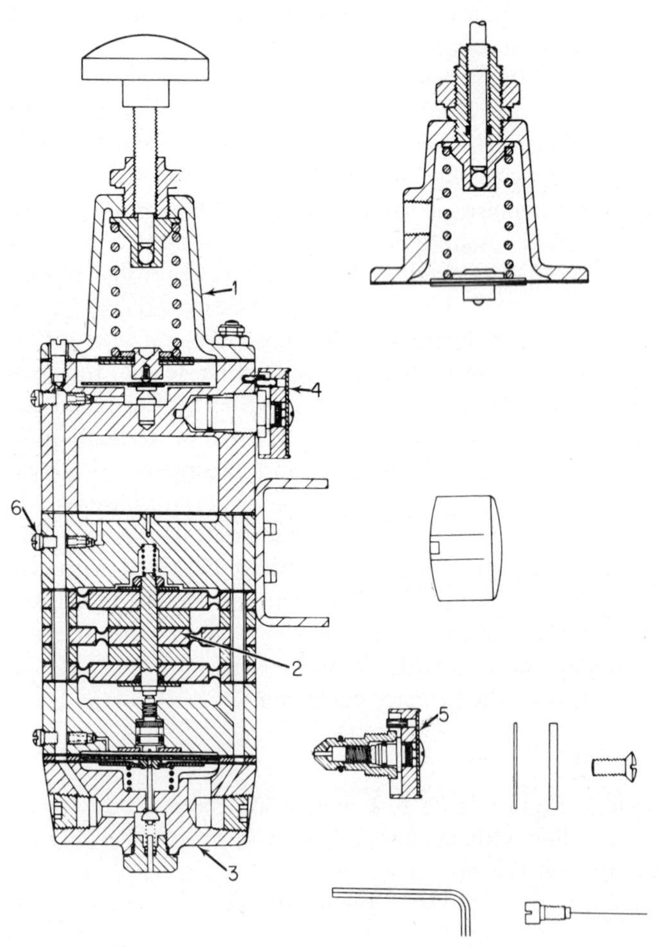

Fig. 11-34 Pneumatic-type controller. (*Moore Products Co.*)

chambers of the diaphragm stack. These two signals pass through a reversing plate located on the outside of the unit. The position of the reversing plate determines the action of the controller. When the plate is positioned so that the set-point signal is applied to the upper section of the control diaphragm and the transmitter signal to the lower section, the controller is direct-acting. Reversing the signals by repositioning the reversing plate makes the controller reverse-acting.

In operation the control point is set to the desired value by adjustment of the control-point adjustment. Since the signal from the measuring transmitter acts to oppose that of the set point, the controller is in an unbalanced state except when the two signal pressures are equal. Any unbalance between the two pressures causes the control-diaphragm stack to move up or down, depending upon which pressure is the greater. Movement of the diaphragm stack alters the rate of air being vented through the nozzle, which changes the pressure above the air-relay diaphragm. For instance, if the stack is made to move down, the nozzle would be closed, causing a pressure increase above the air-relay diaphragm, forcing it down and thereby closing the automatic bleed and opening the pilot valve. This results in an increased air pressure to the control valve.

Since this same pressure is admitted to the lower side of the bottom diaphragm of the control stack, a balance of forces is established when the force created by the output pressure on the stack is equal to the force created by the difference in pressure between the set point and that of the measuring transmitter. The action described has the controller operating as a 2:1 reducing-ratio relay or a proportional controller with a 200 percent proportional-band setting. To improve the action of the controller, a proportional-band adjustment is added in the form of a needle valve which is installed in an air passage between the top of the upper-stack and the bottom of the lower-stack diaphragms. The restricted-supply airflow beneath the reset diaphragm is also admitted to the top of the upper-stack diaphragm through a passage and restriction. Thus, with this system of restrictions, a pressure-dividing arrangement is provided. When the proportional band is adjusted, a portion of the output pressure can be made to act on the upper-stack diaphragm to amplify the action of any unbalance between the set-point and transmitter pressures, thereby making the proportional band theoretically adjustable from 0 to 200 percent.

To further improve the controller, an adjustable reset in the form of a needle valve is added. When this needle valve is adjusted, the rate of airflow in and out of the reset chamber can be altered, thereby changing the rate of reset. Reset action is provided by the pressure within the reset chamber, which acts to increase or decrease the pressure beneath the reset diaphragm. A pressure change beneath the reset diaphragm produces a pressure change in the chamber above the diaphragm stack. This change is proportionate to the proportional-band setting; therefore, reset action is related to proportional-band setting.

The set-point regulator is no more than a precision air-regulating device which maintains its output pressure wtihin ±1 in. of water pressure. The air relay which maintains the controller's output pressure consists of a diaphragm and a valve which functions as an automatic atmospheric bleed and a supply-air valve.

The Moore model 50 is identical to model 55, except that the set point on the model 50 is remotely set, whereas the model 55 set-point regulator is integral with the controller. Model 56 shown in Fig. 11-35

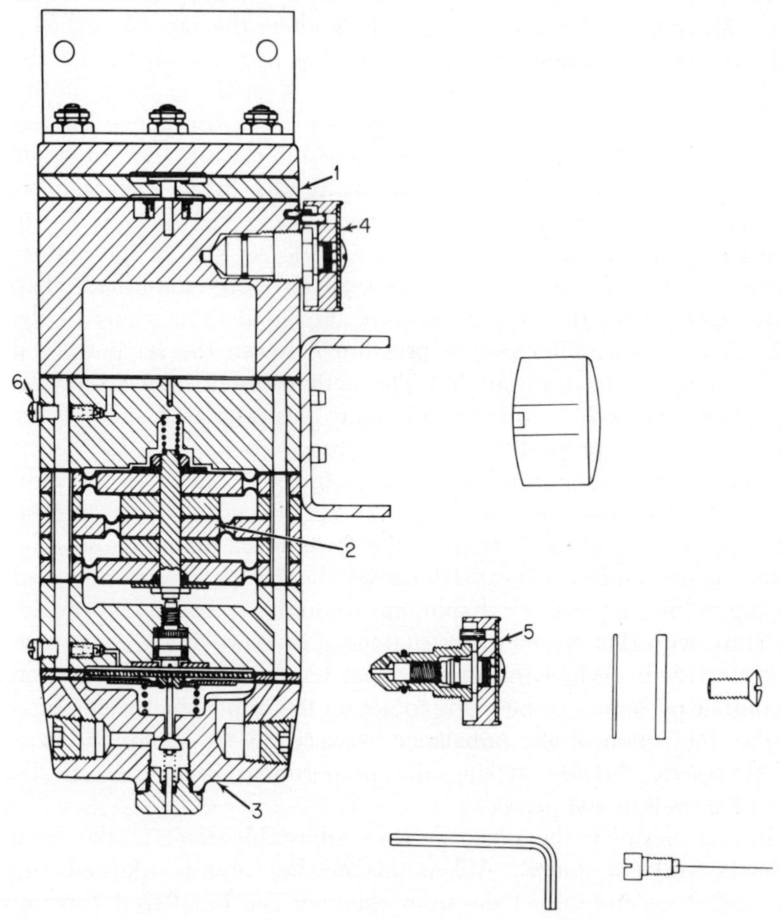

Fig. 11-35 Pneumatic-type controller. (*Moore Products Co.*)

Fig. 11-36 Computer-set Syncro control station. (*Moore Products Co.*)

is identical to the model 50, except that the 56 has built-in cutoff relay 1 for manual-automatic control.

Moore Products Co.

PRINCIPLE OF DESIGN: Figure 11-36 is a Moore Products Co. computer-set Syncro* control station. These are indicating, pneumatic-analog control stations which accept digital control commands directly from a process control computer. All incorporate the unique Syncro concept of procedure-free, instantaneous transfer between control modes. The complete line includes versions in both the 6- by 6-in. panel and 2- by 6-in. (Mini-Syncro) panel sizes shown in Fig. 11-37—four models for supervisory computer control and four for DDC. Operation is as follows:

Supervisory computer control. The computer controls the set points to the analog control loops.

DDC. The computer controls valve loading directly without intermediate analog loops.

Procedure-free, bumpless transfer. Syncro stations may be switched instantly between all control modes without concern for prealignment procedures. A pair of pneumatic Syncros in each control station permits this transfer by automatically maintaining necessary signals in balance and in a state of readiness for instant transfer. All the operator need do is to flip the front-of-station selector switch to change the control mode. The computer connects to the analog loops of the control station through a permanent, magnet-stepping motor which incorporates magnetic memory. Through the memory capability of both the stepping motor and the pneumatic Syncro, smooth transfer back and forth be-

* Registered trade name, Moore Products Co.

tween computer-controlled modes and station-controlled modes is assured.

Complete pneumatic energy source. Except for the stepping motor, computer-set Syncro stations function entirely on a pneumatic supply pressure. Thus, these stations make a sensible backup for the computer, which operates entirely from an electrical source. In the event of an electrical failure, either sustained or momentary, the Syncro stations can take complete control of the process, using the supply air stored in the component's high-pressure air tanks. If electronic transmitters are employed and they go down with the computer, fully pneumatic manual backup is available for an orderly process shutdown. If pneumatic transmitters are employed, automatic and/or manual control can sustain the process operation.

Isolation of control modes. The computer and local control modes are completely isolated from each other. In the computer mode, the station pushbuttons are disconnected so that the operator cannot interfere with the computer. In manual or local/automatic control, the computer input commands to the stepping motor are disconnected so that the computer cannot override the operator.

Feedback-status contacts. All stations incorporate feedback contacts to the computer to inform the computer of the existing control mode. This enables the computer to stop sending ineffectual commands and to print out for supervisors the times and occasions when an operator altered a control mode.

Single-station complete control stations. Complete analog control stations with computer inputs are available in the 6- by 6-in. and 2- by 6-in. models. Models 5217C3 (6×6) and 5227C3 (2×6) are supervisory control stations which permit a computer-directed set point to the integral-analog controller of a single-loop system, local-directed set point to the same controller, and manual control directly to the valve. Models 5218C3 (6×6) and 5228C3 (2×6) are DDC control stations which permit computer control directly to the final control element (valve); single-loop, analog control backup (in which the operator manipulates the set point); and manual control to the valve. All components are integral and no other stations (such as set-point or remote-set cascade types) are required. The stations are continuously self-tracking and thereby allow instant transfer between all operating modes. For both supervisory and DDC stations, automatic analog backup or manual is available instantly.

Full-analog indication. All pneumatic indicators show the full extent of the range of measured-variable, set-point, or valve loading. Measured variable (black pointer) and set point (red pointer) are displayed on the 3½-in.-diameter duplex gauge on 6- by 6-in. stations and on the 4-in. vertical-scale gauge of 2- by 6-in. stations. Valve-loading or station output is displayed on a 2-in. horizontal scale gauge on both stations. Continuous analog indication from a group of stations projects the process and control conditions all at one time. Information, such as how fast and in what direction a variable is changing, how far a variable is from the end of the measured range, and how far off set point a variable is with respect to full scale, is always present.

In 6- by 6-in. stations, duplex scanning gauges are available to align all the process and set-point signals at 12 o'clock high (or any other gauge position). This permits ease of readability or scanning of an off-control process on a large number of stations. On 2- by 6-in. stations, a "day-glo" red pointer "hides" behind the black pointer when on control and appears when an off-control condition arises. On 2- by 6-in. stations, an optional dc alarm light is available for back-lighting the transparent rocker bar.

Operator orientation. For operator convenience, Syncro control stations can be related to the process so that in manual or local/automatic control, the station pushbuttons have the same orientation. It does not matter whether the control valve is air-to-open or air-to-close or what effect the valve action has on the process, depressing the left-hand pushbutton always decreases the controlled variable, while depressing the right-hand pushbutton increases it. This operating simplicity aids the operator who must be familiar with the process control computer as well as with conventional control hardware.

Common-emergency manual control. Any number of stations on line with a computer can be switched to manual control simultaneously by a single pushbutton connected for this purpose; or the switching action may be initiated by the computer itself. Thus, in the event of an electrical or computer failure, the backup manual controls in the Syncro station can take full control without disturbing the process. The stations can be returned to computer control, at any time (without a bump), by deactivation of the original switch action. If the computer is out of service for any length of time, stations equipped with backup analog controls can be switched to local/automatic before the emergency manual switch is returned to normal. The analog controllers then assume automatic control of the process without a control interruption.

Common-emergency local/automatic. On control stations, any number of backup analog control loops (local/automatic mode) can be activated by an external pushbutton or by the computer itself. Since these backup loops are continuously aligned while the computer is in control, instantaneous transfer to the local/automatic modes, without the process being disturbed, is assured. Furthermore, full alignment exists in the inactive computer mode, thus permitting smooth transfer back to this control mode at any time.

Reset bypass controllers. Stations incorporating automatic analog controls are equipped with reset-bypass controllers. These controllers permit instantaneous switching between computer and local/automatic modes and the manual mode without having to wait for the reset pressure to bleed across the reset needle valve. In ordinary controllers, the waiting period, approximately four reset time constants, must be satisfied before switching without a bump is possible. Controllers are either two mode (proportional plus reset) or three mode (proportional plus reset plus rate).

Plug-in identical components. The pneumatic Syncros and the controller incorporate plug-in barbs for quick-disconnect removal and installation. Conventional Syncro stacks used in control stations are identical and may be substituted for each other. Synchro stacks with attached stepping motors (motorized Syncros) are also identical and can be substituted for each other. Even motorized and nonmotorized Syncros can be interchanged if a spare of one type should not be available. Both the controller and the jet synchros are held in place by two ¼-turn, quick-disconnect fasteners. The pneumatic connections include self-sealing O rings which permit removal of inactive instruments without disturbing control from the station. While on manual control, the controller and set-point Syncro can be removed without affecting manual operation.

Built-in end-scale limits. All motorized Syncros employ unique end-scale limits which operate off the relative motion of the range-spring-adjusting mechanism. No electromechanical clutches, limit switches, pressure switches, or jamming members are used. The limits are fixed and function at approximately 2 and 16 psi. These limits ensure equipment safety without interfering with the adjustable program limits furnished by the computer on set-point, process, or controlled variable.

6- by 6-in. stations. The 6- by 6-in. stations employ a 3½-in.-diameter duplex gauge and a horizontal edgewise gauge for all necessary indications.

Selector switches are located at the bottom left and right corners of control stations. Loading stations have a single selector switch at the bottom right-hand corner.

Pushbuttons for manual control are located to the left and right of the valve gauge. Through operator orientation, depressing the right-hand pushbutton on control stations always raises the process variable, while depressing the left-hand pushbutton lowers the variable.

Access to the internal working components is made through a side-opening front door in which the 3½-in.-diameter duplex gauge is mounted. If the gauge is a scanning type, a knurled knob on the rear of the gauge is used for rotating the gauge dial to the desired position.

The major components of the station are mounted on an assembly block affixed to the rear of the station. Each control station incorporates two pneumatic Syncros plugged into this block. One of the Syncros (depending upon whether the station is for supervisory or DDC control) includes the stepping motor. The combination of a Syncro and stepping motor is referred to as a motorized Syncro; its function is to convert digital commands from the computer into a 3- to 15-psi pneumatic output.

The controller plugs into a manifold mounted on the outside of the rear block. Pneumatic connections are made through ⅛-in. NPT pipe taps on the manifold. If a field-mounted controller is employed, the manifold is deleted in favor of ⅛-in. NPT pneumatic connections. Electrical connections are made through a ½-in. NPT conduit connection on the rear block.

Loading stations utilize a motorized Syncro only, which simply converts the digital input commands into a pneumatic set-point or valve-loading pressure.

Switching components on all stations include: stack-diaphragm transfer switches, which perform all necessary instrument signal switching; rotary selector valves, which produce all supply pressure switching; and electrical microswitches, which provide motor interrupt and mode feedback functions with the computer. All microswitches are SPDT, long-life type.

2- by 6-in. stations. The 2- by 6-in. station, shown in Fig. 11-37, employs a 4-in. vertical scale, duplex gauge for indication of process and set point. On control stations, the gauge displays the process variable on the right black pointer and set point on the left red pointer. On set-point loading stations, a single-pointer gauge indicates the station output on the right black pointer. On valve-loading stations, the right black pointer indicates process, while the horizontal gauge, beneath the vertical gauge, indicates valve-loading pressure. The horizontal gauge also displays valve pressure on all the control stations.

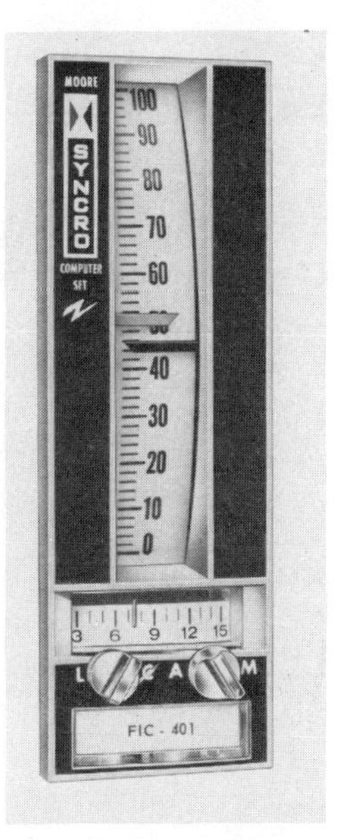

Fig. 11-37 Mini-Syncro. (*Moore Products Co.*)

Selector switches are located under the horizontal valve gauge. Loading stations have a single selector in the center of the same area. The switches are designed for finger-tip operation, and are raised from the face of the station to permit ease of access on the high-density station. Pushbuttons for manual control are located behind the identification rocker bar at the bottom edge. The rocker bar is pivoted at the center, allowing the operator to depress the left or right side to lower or raise the process variable or station output, respectively. A socket screw along the bottom of the station bezel holds the transparent rocker bar in place. This permits the rocker bar to be removed, with the station installed, for placing an identification nameplate within a pocket of the bar. An optional alarm light is mounted behind the rocker bar.

The 2- by 6-in. control station consists of two modules: a small, lightweight panel module containing all the essential operator controls and indicators; and a remote module containing the Syncro stacks and controller shown in Fig. 11-38. A flat ribbon-style, multiple tubing connects the two modules. The remote module is rack- or wall-mounted with the essential hardware being completely accessible from one side. The components of the remote module are exactly the same as those used on the rear block of the 6- by 6-in. control station as shown in Fig. 11-36. Electrical connections on both the panel module and the remote module are made at integral screw terminal strips.

Loading stations use a single motorized Syncro, plugged into the rear

Fig. 11-38 Plug-in type Syncros. (*Moore Products Co.*)

of the 2- by 6-in. panel module. To attain high-density packaging, adjacent motorized Syncros are honeycombed top and bottom. Pneumatic connections are made at the rear of the panel module. Electrical connections are made on a no. 6 screw terminal strip attached to the motorized Syncro. No remote module is used with the loading stations.

Moore Products Co.

PRINCIPLE OF DESIGN: Shown in Fig. 11-39 is a simplified diagram of the Moore Products Co. motorized Syncro. The motorized Syncro is the key difference between conventional and computer-set Syncro control stations.

The stepping motor connects to the pneumatic mechanism through an internally threaded bushing. This bushing engages the threaded lead screw of the pneumatic turbine wheel, which in turn varies the range-spring loading on the regulator. In the computer operating mode, the lever arm of the pneumatic brake constrains the turbine wheel so that the wheel cannot turn. The turbine wheel can, however, move axially. Thus, as the motor rotates its affixed bushing, the lead screw and turbine-wheel assembly moves in and out of the bushing and varies the spring-loading force on the regulator.

In the station operating modes (local/automatic and manual), the pneumatic brake disengages the lever arm from the turbine wheel. The wheel is now free to rotate and thread itself in and out of the motor bushing. In these station modes, the stepping motor power is disconnected and the motor bushing becomes a fixed member, held in place by the inherent magnetic detent of the motor. At the up-and-down nozzles, pushbutton or deviation-controller pressure then impacts air

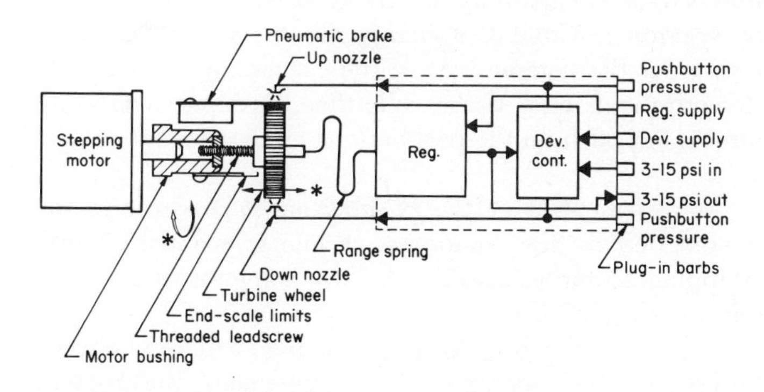

Fig. 11-39 Simplified diagram of a motorized Syncro. (*Moore Products Co.*)

against the turbine wheel, rotating it in the desired direction to vary the spring loading on the regulator.

The stepping motor and pneumatic nozzles act to reposition the same turbine wheel and, resultantly, the same regulator range spring. This common function assures smooth transfer between computer and local control modes instantaneously without prealignment and without disturbing the process.

The built-in end-scale limits employ: (1) a bracket fastened to the motor bushing and (2) a washer attached to the turbine wheel. The washer includes a small tip which revolves inside the limit gap of the bracket. The gap is sized so that, at either limit, the side of the washer tip positively engages the side of the bracket without jamming of the separate members. In the computer mode, the turbine wheel moves axially until the limits engage and then rotates against the friction of the brake arm along with the motor bushing. No further axial motion in that direction occurs. In the local station modes, the turbine wheel simply rotates until the washer engages the bracket on the stationary motor bushing. The force of impacting air from the nozzles is insufficient to turn the motor bushing. The end-scale limits are set at nominal values of 2 and 16 psi.

The pneumatic regulator converts the forces exerted on the range spring by the turbine wheel into a 3- to 15-psi output pressure. The deviation controller aligns the regulator output with an independent input whenever supply pressure is connected to this section. A pressure difference between output and input generates control air pressure to the up-and-down nozzles, which then rotate the turbine wheel to vary spring loading on the regulator. Thus, regulator output pressure automatically tracks the independent input. This servo operation is essential to the procedureless transfer capability of Syncro control stations.

OPERATION OF STATIONS: *Computer modes.* In the computer mode, digital commands from the computer vary the pneumatic output from the motorized Syncro regulator. At the same time, pushbutton pressure for manual control is cut off from the pushbuttons by the computer/local selector switch.

Local mode. Manual control at the pushbuttons takes over in the last set position dictated by the computer. While in manual control, power to the stepping motor is disconnected by a microswitch which works off the selector switch.

There is never any "bump" when the unit is being switched between operating modes because the stepping motor and pneumatic pushbuttons each take over operation of the Syncro stack where the other leaves off.

Feedback contacts to the computer, which work off the computer/local selector, deliver closed- or open-circuit connections back to the computer.

SUPERVISORY-TYPE STATION: *Computer-automatic mode.* The computer varies the controller set point by actuating the motorized set-point Syncro. In accordance with these commands, the controller automatically varies the valve pressure to hold the process variable at the set point. The operator cannot change the set point because the pushbuttons are isolated by the selector switches. In this mode, the deviation controller in the valve-loading Syncro receives supply air. This permits the valve-loading Syncro to continuously align its regulator output with controller output pressure, always ready to assume direct control of the valve in the manual mode. In the automatic mode, the valve-loading Syncro output is blocked by the selector switches.

Local/automatic mode. The only difference between the computer and local/automatic control modes is that the pushbuttons on the front of the station are activated in local/automatic, and they become the sole means of adjusting the set point. The computer, in the meantime, is disconnected from the stepping motor by means of the microswitch mounted on the computer/local switch. Thus, transfer to local/automatic is made without creating any disturbance.

The computer takes over operation of the set-point Syncro wherever it was left by pushbutton adjustment or by the self-synchronizing system. The station can be switched back to computer/automatic at any instant without a bump.

Manual mode. When switched to manual through the singular and instantaneous transfer of the Auto/Man selector, several station functions occur. First, the stored output of the valve-loading Syncro, which has been following valve pressure, passes to the valve. Supply air from the station shuts off the controller output through a built-in pneumatic cutoff relay in the controller. Supply air to the deviation controller of the valve-loading Syncro is cut off, making the jet-Syncro responsive to pushbutton operation only. The operator then may raise or lower the valve pressure by using the station pushbuttons. The deviation controller in the motorized set-point Syncro now receives supply air, causing the jet Syncro to automatically track the process variable.

In addition to shutting off the controller output, the controller cutoff relay connects valve pressure, as generated by the valve-loading Syncro, directly to the controller reset chamber. Since process and set point are matched by the set-point Syncro, the shut-off controller output then equals the feedback applied to the reset chamber. Also, since feedback bypasses the reset needle valve, the controller output tracks feedback valve loading without delays. The controller is aligned to valve pressure and is always ready to resume control of the valve. Transfer can be made to computer or to local/automatic without disturbing the process. Feedback contacts to the computer produce a closed-contact sense in the computer/automatic mode, and an open-contact sense in either local/automatic or manual.

DDC MANUAL BACK-UP STATION TYPE: *Computer mode.* In the computer mode, digital commands from the computer vary the pneumatic output from the motorized regulator to the control valve. At the same time, manual control through the pushbuttons is cut off by the computer/manual selector switch.

Manual mode. The operator manipulates the output of the Syncro by depressing the pushbuttons. A microswitch, operated by the computer/manual selector, disconnects the computer commands to the stepping motor by opening common lead on the motor. There is never any "bump" when switching between operating modes because the stepping motor and pneumatic pushbuttons each take over operation of the Syncro where the other leaves off.

As previously stated, the process variable is displayed at all times on the station gauge, 3½-in.-diameter gauge, 6- by 6-in. stations; 4-in. vertical gauge, 2- by 6-in. stations. In either control mode, the gauge serves to indicate clearly the position of the process variable. Such indication in a DDC system is extremely valuable to an operator, providing him with a convenient, quick check of the overall system. In the manual control mode, it enables him to make intelligent variations in valve pressure.

Similar to the manual back-up station, the DDC control station also serves as a full-time digital-to-analog converter linking the computer to the control valve. It furnishes convenient and continuous analog indication of the state of the process variable and the control-valve pressure, along with the local set point (when used). This capability eliminates the need for multiplexed D/A converters.

Moore Products Co.

PRINCIPLE OF DESIGN: The synchronous motor used in the Moore Products Co. motorized Syncro regulators and transmitters is a three-lead, bidirectional motor that can be used as either a dc stepping motor or an ac constant speed motor.

Basically, the motor consists of a permanent magnet rotor and a two-phase electromagnetic stator. Interaction between the alternating magnetic field generated in the stator and the unidirectional flux of the permanent magnet produces the necessary torque and rotation.

A 60-Hz ac signal runs the motor at a synchronous speed of 72 rpm. A single-pole, three-position switch provides complete forward, reverse, and off control.

Shown in Fig. 11-39 is a single-phase hookup. When the unit is operated from a single-phase source, a phase-shifting network consisting of a resistor and a capacitor must be used. A 500-ohm, 5-watt resistor and a 0.75-mfd, 330-volt ac capacitor provide satisfactory operation at any frequency between 50 and 60 Hz. Other frequencies may require different resistor and capacitor values to give the desired 90° phase shift.

When operated from a two-phase source, the motor needs no shifting network. However, at frequencies other than 50 and 60 Hz, the voltage should be adjusted to maintain the current at its rated value (0.07 amp, normal; 0.1 amp, maximum).

Variable speed control can be achieved with a two-phase, variable-frequency power source. The motor-shaft speed in revolutions per minute is equal to 1.2 times the applied frequency in cycles per second.

When used for phase-switched dc stepping, the motor can be connected to provide either 100 or 200 steps per revolution.

The motor can be used with a center-tapped dc supply and single-pole, double-throw switching or with a single-ended dc power source and single-pole, double-throw switches to provide double-pole, double-throw switching.

Switching can be accomplished mechanically with SPDT switches, with mercury-wetted relays (break before make), or through a commutator-brash arrangement. Electronically, switching can be performed by transistors or tubes.

A dc square wave makes the motor run at distinct interrupted steps. This on-off-type rotation is compatible with digital logic and produces resolution which is strictly defined by discrete steps. The motor can step at varying speeds and can start, stop, and reverse direction without losing any steps.

One hundred steps per revolution is the more common stepping configuration and can be used on a computer's digital-output channel. The stepping rate can be any value up to 100 steps per sec. The motor takes 1,000 steps to drive the motorized Syncro regulator or transmitter output through a full-scale 3- to 15-psi change.

A phase-shifting capacitor is required for proper stepping and direction control. The value of this capacitor is 1 mfd ± 6 percent at 330 volts ac.

Transient voltages are generated during dc stepping as voltage is switched across the windings. Unless some means of limiting or removing these voltages is provided, they could cause faulty operation or damage to the motor or to circuit components. The most common method of suppressing transient voltages employs shunting diodes, which is accomplished by the circuit end.

Taylor Instrument Companies

PRINCIPLE OF DESIGN: Because reducing the size of the control housing and the instrument panels saves cost to industry, most all instrument factories have reduced the size of their instruments without sacrificing accuracy. Shown in Fig. 11-40 is a good example of what the manufacturers are doing to offset this undesirable feature. The instrument, model 640R, is designed to fit in one or more 3- by 6-in. Quick-Scan* cases. This instrument is a servo-powered deviation indicator. The design makes it possible to take the meter out of service and remove the indicator without disturbing the control mechanism which is equipped with a small glass ball within a glass tube that shows the instrument is in balance and can be changed from the Auto to the Manual position.

The input and output of this controller are 3 to 15 psi. Recommended minimum and maximum temperatures of the instrument should not be less than 30 or greater than 140°F.

Three styles of indication scales are used: manual, servo-driven, and fixed scale. The method used to separate the indicating and the control mechanisms is due to the case design. The top section can be disconnected and pulled out enough so that the indicating mechanism can be removed and sent to the shop for repairs; or in some cases repairs can be made at the job sight.

* Registered trade name, Taylor Instrument Companies.

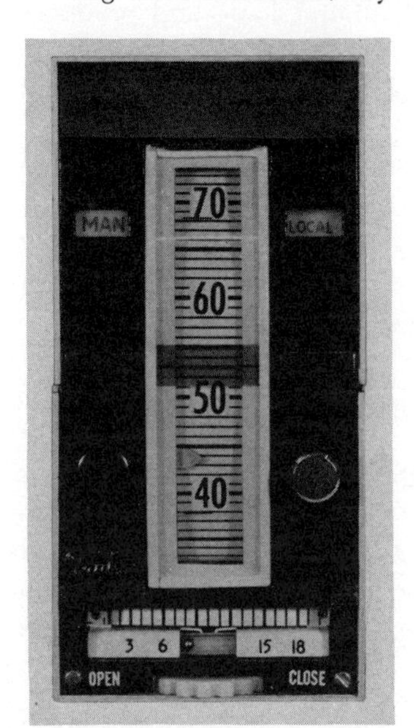

Fig. 11-40 3- by 6-in. Quick-Scan control. (*Taylor Instrument Companies*)

The tape-scale arrangements are as follows:

- Manual: A 9-in. tape, conventional form, is set manually by a knob located on the front of the instrument.
- Pneumatic servo-driven: This pointer indicates the controller set point whether on local or remote set.
- Fixed scale: The scale in this case is the simplest form of scale. The pointer on this scale reads ±20 percent deviation from the set point.

The manual-set and pneumatic servo-driven forms are unique in that the tape is movable by the set-point adjusting knob or by the pneumatic servo in the cascade form. Thus positioning of the tape establishes the required set point behind the green scan-band hairline index. If a difference exists between the set point and process variable, the deviation-indicating pointer moves away from the green scan-band and shows its normal red color. When no difference exists, the deviation-indicating pointer is at the center of the tape scale and appears dark brown in color when viewed through the green scan-band. The deviation pointer also indicates the process-variable value when read against the calibrated tape, as long as the deviation is no greater than 20 percent of the tape-scale range.

The fixed-scale form is an economical variation in which the tape is not moved and permits indication of the deviation value but not the process variable.

Bumpless automatic to manual transfer is accomplished by means of a two-position Auto/Manual switch and the no-flow ball gauge which indicates whether a differential exists between the controller and the manual output. When the set-point adjusting knob (in manual) or the manual output adjusting wheel (in automatic) is manipulated, the ball of the gauge is centered before the position of the Auto/Manual switch is changed, and bumpless transfer is easily effected.

Controller responses, including all combinations of proportional, reset, and Pre-act,* are available. Changing action is simply effected by loosening a lock screw and turning the action-reversing plate 90°. The reversing plate is located on the manual slide (not on the controller) so that even if the controller is replaced, instrument action cannot be accidentally reversed.

Figure 11-41 is the control mechanism used in the deviation instrument to make it a complete three-mode controller. All the dials are well marked and easy to adjust. The control can be made direct or reverse, which allows it to be flexible enough to be changed from one location to another where its operation can be adjusted to the new process. Note that the control unit is built so that various components can be changed

* Registered trade name, Taylor Instrument Companies.

Fig. 11-41 Quick-Scan control unit. (*Taylor Instrument Companies*)

or repaired without disburbing other parts of the mechanism. Flexible tubing is installed on other similar controllers without any repiping.

The design of the 640R controller is rugged in its construction and simple in its operation and maintenance.

Taylor Instrument Companies*

PRINCIPLE OF DESIGN: Figure 11-42 is a cutaway view of Taylor Instrument Companies' plug-in-type pncumatic controller, which the manufacturer refers to as its Transcope† controller.

The design of the controller permits it to be plugged into a manifold

* Carroll, *Industrial Instrument Servicing Handbook*, pp. 11-36, 11-37.

† Registered trade name, Taylor Instrument Companies.

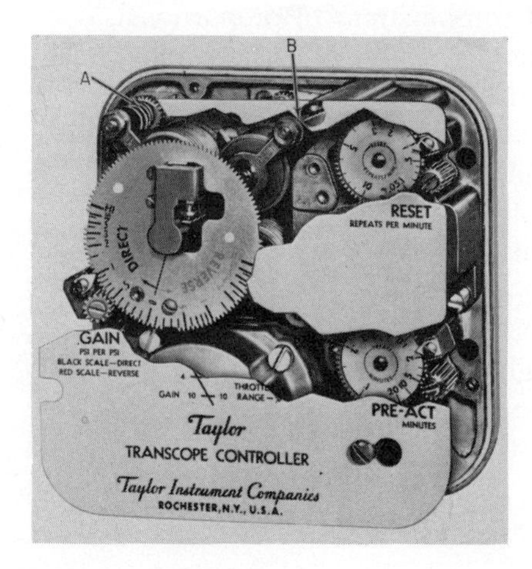

Fig. 11-42 Plug-in-type pneumatic control. (*Taylor Instrument Companies*)

on Taylor's recording or indicating receivers, or it may be installed on a field-mounted manifold.

The controller is based on an adaptation of the motion-balance principle, using bellows as the pressure-sensitive elements. The four bellows shown schematically in Fig. 11-43 act against a force plate which pivots about a wire in tension. The rocking motion of the force plate about this universal pivot is restrained by two range springs. A drive pin rigidly attached to the plate transmits motion of the baffle linkage, which regulates nozzle closure. The nozzle back pressure operates a direct-acting relay to supply the feedback and the output circuit.

A simple proportional controller is made by removing the Pre-Act valve and the reset bellows. The gain of the instrument is then adjusted by rotating the baffle assembly about the center line of the nozzle. When the controller is in a balanced state, the nozzle and drive pin have a common center line. If the baffle assembly is placed directly over the process bellows, the controller is in direct-action position and

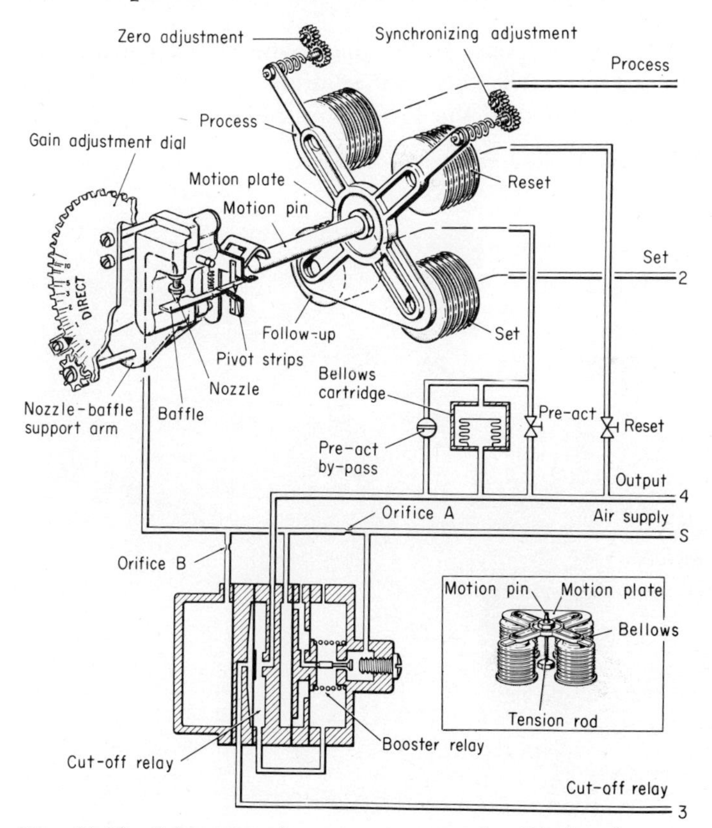

Fig. 11-43 Schematic of pneumatic control. (*Taylor Instrument Companies*)

has maximum gain. An increase in the process pressure moves the drive pin away from the baffle assembly, which caps the nozzle and increases the output pressure. The increased output pressure causes the follow-up bellows to expand and rock the force plate. However, this follow-up motion does not move the baffle because the resulting drive-pin motion is parallel to the face of the baffle assembly.

If the baffle assembly is placed halfway between the process and reset-bellows corner, the controller is direct-acting with a gain of 1. In this position, the motion of the drive pin, due to either the process or follow-up bellows, acts at an angle of 45° to the face of the baffle linkage. The instrument then comes to balance when the motions due to the process and follow-up pressures are equal. When the baffle assembly is placed over the reset corner of the force plate, the motion due to the process bellows is parallel to the face of the baffle assembly, and the gain is zero. The controller operates in reverse action as the baffle assembly is rotated through the quadrant between the reset and set bellows.

The reset and Pre-Act response in the pneumatic circuit corresponds to that in the Taylor series 100 Fulscope* controller. A Pre-Act bypass bellows assembly is included in the Pre-Act instrument to improve the

* Registered trade name, Taylor Instrument Companies.

Fig. 11-44 Quick-Scan recorder. (*Taylor Instrument Companies*)

Fig. 11-45 Bank of Quick-Scan controllers and recorder. (*Taylor Instrument Companies*)

stability of the controller. It allows a partial feedback or transfer of output pressure to the follow-up bellows for all settings of the Pre-Act needle valve and substantially reduces the effect on the output pressure of sudden supply-pressure changes and other transient disturbances.

Taylor Instrument Companies

PRINCIPLE OF DESIGN: The Taylor instrument shown in Fig. 11-44 has many desirable features such as a powerful pen-drive mechanism and two clocks which provide a wide range of chart speed. The recorder as shown can be partially withdrawn while the instrument is in operation without disturbing the process. This feature allows those in charge of production of plants to check on operators for several hours past. The recorder can be designed with two recording pens for processes. In addition to recording process variables, provisions are available for installing alarm contacts. Note that in Fig. 11-44 the entire instrument mechanism is mounted on a sliding base so that the instrument can be withdrawn from its case and returned to the instrument laboratory if major repairs become necessary.

Taylor Instrument Companies

PRINCIPLE OF DESIGN: Shown in Fig. 11-45 are five Taylor Instrument Companies Quick-Scan controllers and one recorder installed in one case. Such installations provide means for plugging into the recorder any two of the five process variables being controlled by the controllers. The controllers and recorder are constructed in such a manner that they can be switched from automatic to manual or vice versa.

An output indicator on each controller has a glass ball located in the center of the output meter. When a switch is made from automatic to manual or vice versa, the pressure to the control valve is adjusted until the ball is exactly in the center of its glass containing tube. The switch then can be made without disturbing the final controlling device (this is usually a control valve); this action provides a bumpless transfer. The controllers can be equipped with adjustable proportional-band, proportional-plus-reset, or proportional-plus-reset-plus-rate action (Pre-Act).

CHAPTER TWELVE

Electrical and Electronic Controllers

IN MODERN INDUSTRY electrical- and electronic-measuring and -controlling equipment has become very important for controlling some of the complex processes where time lags, computers, etc., are used. However, just as a complex continuous process is a combination of several simple controllable units, so a complex control mechanism involving correction for several variables is essentially an interrelated group of simple mechanisms. Computers designed to receive electronic signals especially account for the tremendous increase in the use of electronic instruments. If pneumatic controllers are used, the extra expense of transducing from pneumatic to electronic can become very expensive. In addition, more equipment is involved, which increases maintenance cost, and better-trained technicians are needed to service such a system.

Description of Electrical Control Instruments

ALNOR INSTRUMENT COMPANY

PRINCIPLE OF DESIGN: The Alnor Pyrotroller is a combination of an indicating pyrometer and a simplified electronic control (see Fig. 12-1).

The Pyrotroller operating parts are few in number and of a high order of reliability. Each component is oversized for its required performance to permit operation well below rated capacity. This engineering philosophy results in long, trouble-free field life for the Alnor Pyrotroller.

A single electron tube operates the relay and is responsible to the command of the temperature indicator. This tube performs as an oscillator which can be stopped by the use of sufficient negative feedback.

When the temperature is below the set point, the negative-feedback path is uninterrupted, and the electron tube passes a large amount of plate current, causing the relay to energize the load circuit.

When the negative-feedback path is interrupted by a small aluminum loop on the indicating pointer and the pointer reaches the set point, the electron tube oscillates and will not pass appreciable plate current. When this occurs the relay drops out and opens the load circuit. In the event of power failure to the instrument, the relay drops out and opens the load circuit.

In the event that a thermocouple or extension wire breaks or burns out, the Pyrotrollers with a range of 0 to 800°F and higher react as though an excess temperature had occurred. No extra wires or connections are necessary for this function.

The Pyrotroller has two knobs on the front. One is used for setting the set point; the other is merely a blank and does nothing on the *N*-14. The small screw in the center of the instrument is to adjust the cold-end compensator which is in reality a zero adjustment. The right-hand light indicates the position of the internal relay, and the left-hand light indicates that the instrument is being powered. The instrument can be proportional off and on and can be used as an indicating alarming device.

Fig. 12-1 A combination indicating pyrometer controller. (*Alnor Instrument Company*)

The use of the Pyrotroller is wide and can be incorporated into sophisticated control systems. Maintenance is simple, which is an advantage where a limited amount of skilled labor is not available.

Bailey Meter Company

PRINCIPLE OF DESIGN: Figure 12-2 is one of the components which Bailey Meter Company has used to integrate the operator into the 721 electronic control system. With this unit the operator and the system are continuously communicating with each other. One of the advantages of this unit is that the control room is simplified when information display and control functions are operator-oriented. The control center is designed to permit the operator to supervise intelligently, safely, and economically all phases of plant operation under all conditions.

In the 721 system two types of hand-automatic stations are available: automatic balance transfer and manual balance transfer.

The automatic-balance, hand-automatic station is designed for bumpless transfer from automatic to manual or vice versa. This station is always in use during a power failure, which results in the control drive remaining in its last position until power is restored. Other features

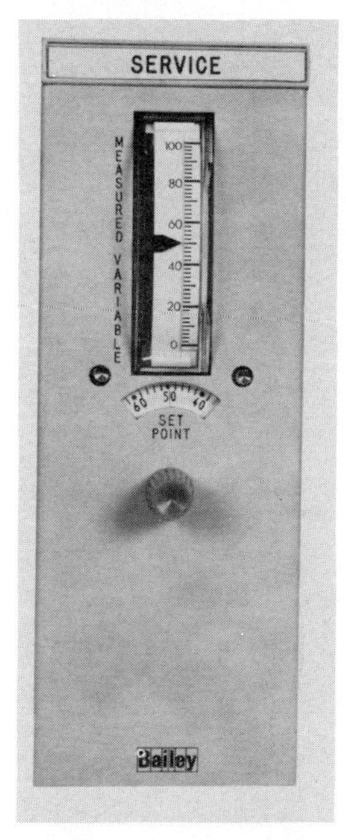

Fig. 12-2 Hand-automatic set station. (*Bailey Meter Company*)

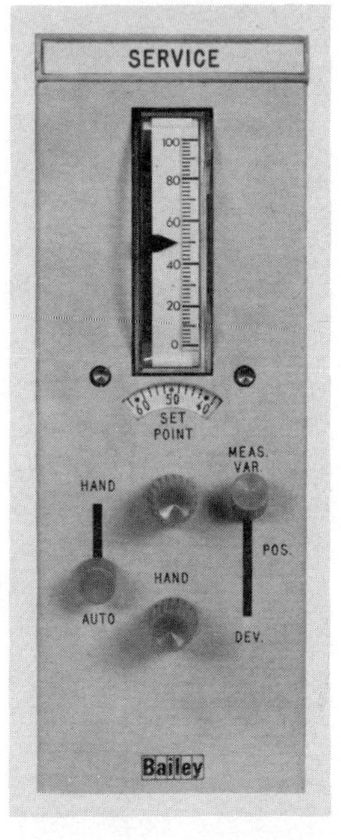

Fig. 12-3 Secondary hand-automatic set station. (*Bailey Meter Company*)

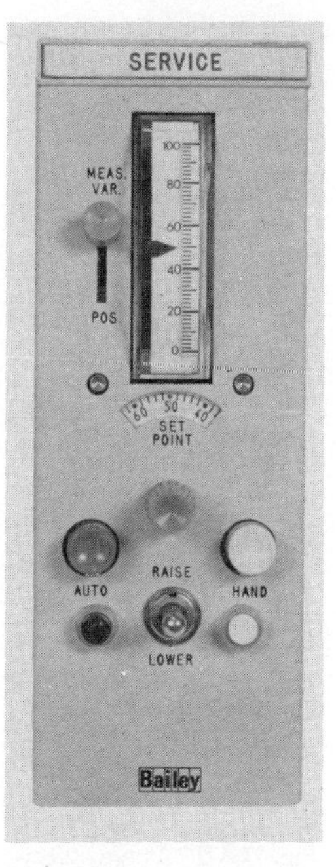

Fig. 12-4 Hand-automatic biasing station. (*Bailey Meter Company*)

are designed to work with the station when a more sophisticated control system such as computers is used.

The manual control is no more than a toggle switch which operates the final control device in a jogging action. The toggle switch is spring-loaded to return to neutral when not being manipulated by the operator. The jogging switch is directly wired to the electrical control drive so that, in the event of control-system power failure, the drive can still be manipulated if power is available at the drive. A station of this nature is shown in Fig. 12-3.

The manual station shown in Fig. 12-4 develops an electrical signal which is manually adjustable to modify signals or to position remote instruments, action units, or power devices.

Figure 12-5 is a Bailey pushbutton and vertical indicator module which is an advanced concept in control and indicating stations. The back-lighted, pushbutton stations may be used for manual-automatic transfer of analog and digital subloops, annunciator functions, and status indications. The vertical indicators provide continuous readings of important plant variables.

Figure 12-6 is an electronic miniature strip-chart recorder used to

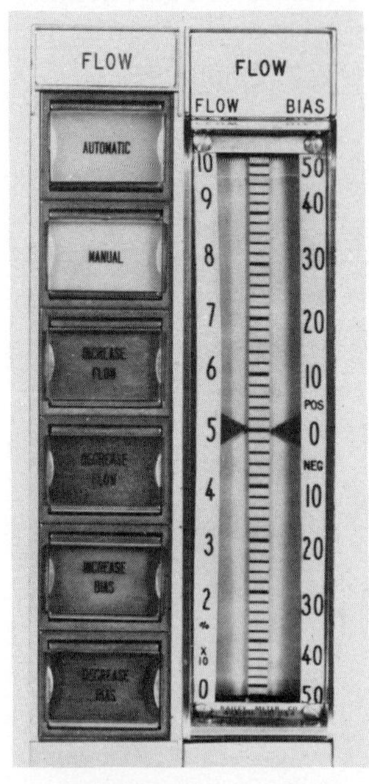

Fig. 12-5 Push-button manual/automatic and vertical indicator control for subloop stations. (*Bailey Meter Company*)

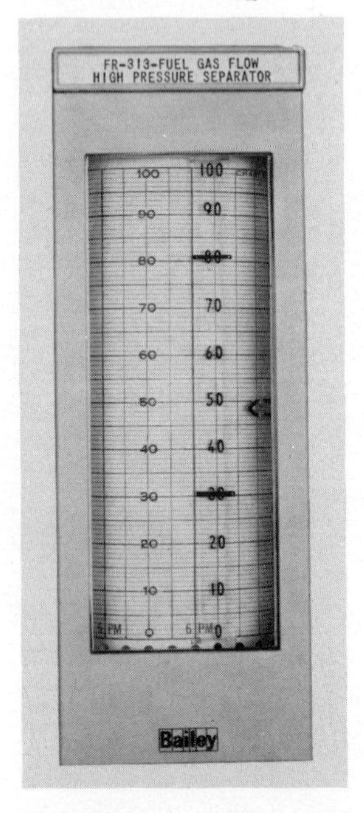

Fig. 12-6 Electronic minature strip-chart recorder. (*Bailey Meter Company*)

record measured variables with a minimum panel space. These recorders provide the operator with the necessary flexibility required to control the console so that the information display and control functions are operator-oriented.

The Bristol Company

PRINCIPLE OF DESIGN: The Bristol electronic pyrometer controller, shown in Fig. 12-7, consists of two separate units: the pyrometer unit, which measures and indicates on a scale the temperature under control; and the electronic Free-Vane* control unit, which opens and closes a relay in the load circuit that regulates the control mechanism. There is no mechanical or electrical connection between the two. The measuring element is not required to do any work in causing the control system to operate and is, therefore, free to measure accurately the temperature in question without interference from the control system.

The pyrometer unit uses the Bristol millivoltmeter measuring mechanism equipped with cold-junction compensation and neutralizer. It is

* Registered trade name, The Bristol Company.

Fig. 12-7 Electronic Pyrometer. (*The Bristol Company, Division of American Chain & Cable Company*)

extremely accurate and sensitive. It is capable of withstanding vibration and the rough operating conditions present in industrial plants.

A lightweight vane, attached to the pyrometer pointer, passes freely between two coils mounted on the control-point spotter arm. These coils are connected into the electronic circuit. The circuit is so arranged that the passage of the vane between the coils alters the electrical characteristics and thereby varies the frequency of oscillation. This change in frequency is used to actuate the control relay.

A pyrometer using a high-torque, rugged, jeweled, millivoltmeter mechanism, is used in Bristol series 536 electronic indicating controllers. The mechanism has 100 ohms internal resistance. Total loop resistance cannot exceed 15 ohms per mv. Minimum span is 10 mv. It is equipped with automatic cold-junction compensation and a neutralizing resistance to provide compensation for changes in coil resistance with ambient temperature. All ranges are critically damped or overdamped.

Operation of the control system does not retard the pointer of the measuring mechanism. There is no reaction effect on the vane at the time of control action. The pyrometer is free to indicate accurately the temperature under control.

The Bristol Company

PRINCIPLE OF DESIGN: Figure 12-8 is a front view of a Bristol electronic controller, model 820, which compares a process-variable to a set-point

Fig. 12-8 Front view of an electronic controller. (*The Bristol Company, Division of American Chain & Cable Company*)

signal. The difference between the two (error signal) is amplified and characterized by proportional-band, reset, and derivative circuits to provide a 4- to 20-ma dc output signal. This signal is used to maintain the process variable at the desired set point by operating the final control device. This device is usually an electric, a hydraulic, or electropneumatic valve; a silicon-controlled rectifier; or a magnetic amplifier and saturable-core reactor, which is shown in the circuit diagram of Fig. 12-9.

The input to the controller is a 4- to 20-ma signal across a 250-ohm resistor, or any other signal and resistance combination to produce a 1- to 5-volt dc input. Note that in Fig. 12-9 the reference voltage is obtained by full-wave rectification of the ac supply, which is then filtered and regulated by a zener-diode circuit to produce a regulated 1- to 5-volt dc set-point signal. The difference between the 1- to 5-volt measured variable input and the 1- to 5-volt set-point signal is a 0- to ± 4-volt dc signal which appears on the zero-centered deviation indicator.

Input to the modulator is the difference between the signal appearing across the deviation indicator and the feedback signal; this is on the order of 0 to ± 4 mv. This combined signal modulates a 100-kc carrier, generated by the oscillator circuit.

The modulator consists of a bridge circuit made up of the center-tapped secondary of the modulator transformer and two solid-state diodes. The 0- to ± 4-mv signal impressed on this bridge circuit causes the diodes to act as variable capacitors, thereby unbalancing the bridge to modulate the 60- to 80-mv ac 100-kc carrier.

The purpose of the input transformer is to isolate the input from the output of the controller, thereby providing high common-mode rejection.

The primary amplifier is a solid-state device that uses two transistors in a common emitter circuit to obtain two stages of voltage amplification.

Operation of the oscillator is such that any signal appearing in the base-emitter circuit is amplified and appears at the primary of the oscillator transformer. A portion of this voltage is transformed by one of the secondaries and produces positive feedback, which is again amplified by the power amplifier and appears at the same transformer secondary. This cycle repeats, causing the oscillation to continue to increase until it stabilizes at the value required by the modulation circuit. Stabilization at the required value is obtained by the design of the oscillator output load which consists of the power amplifier.

The power-amplifier circuit consists of a single transistor amplifier that functions to finally amplify the signal to the 4- to 20-ma dc level required for the output.

A 4-to-1 step-up in the feedback signal is provided by the transformer

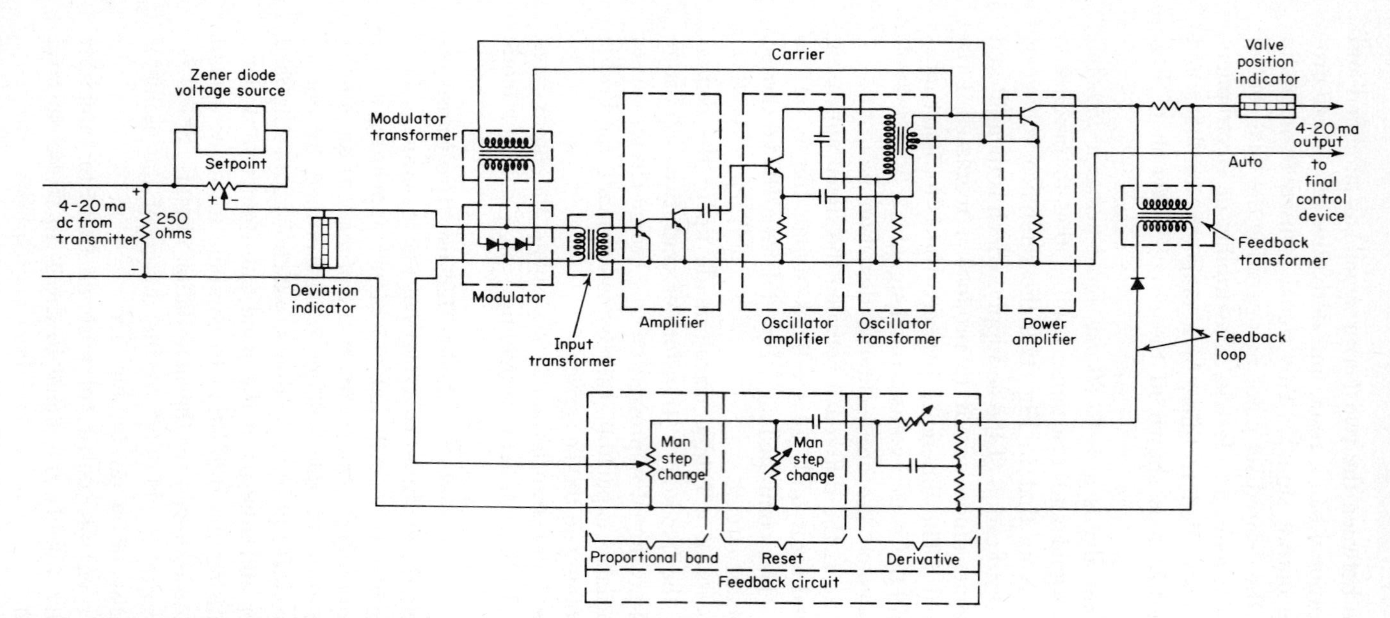

Fig. 12-9 Simplified circuit diagram. (*The Bristol Company, Division of American Chain & Cable Company*)

which also helps provide isolation between input and output of the controller for high common-mode rejection.

The feedback circuit performs the functions of the differentiation (derivative), integration (reset), and control of proportional band or gain.

The feedback signal is amplified 4 to 1 by the feedback transformer, which is necessary to provide overall controller proportional-band settings greater than 100 percent up to a maximum of 400 percent.

Zero frequency gain, sometimes referred to as steady-state or static gain, gives the value of the forward gain of the amplifier without feedback. It is the ratio of an output change to an input change after steady state has been reached.

Change in output of the forward amplifier, proportional to the measured variable step-change in input, is stepped up by the feedback transformer and integrated by the integrating circuit which, since it is in the negative feedback loop, produces differentiation of the controller output. Upon a step-change in input of an integrating circuit, the output is delayed, depending upon the resistance-capacitor time-constant of the circuit. This delay, appearing in the feedback loop, slows the effect of the feedback circuitry, thereby producing an instantaneous large output change from the controller, which gradually decays at a rate proportional to the setting of the derivative resistor.

At the input to the derivative circuit, the divider network limits the maximum derivative gain produced by this circuit. This divider ratio determines the derivative gain and is a compromise between the desired high derivative gain and low high-frequency-noise amplification. The standard series 820 controller has a derivative gain of 20. This means that with a step-change at the input of the derivative circuit, the output will be 20 times this step and will gradually decay at a rate determined by the derivative setting as the output of the derivative circuit gradually increases, thereby increasing the negative feedback and decreasing the controller output.

A step input to the reset network, located in the negative-feedback loop, would produce an opposite response in the controller output, i.e., a gradually changing output until the input is constant. Therefore, the capacitor acts as an open circuit to the constant voltage, and no feedback occurs for steady-state conditions.

A change in feedback at the proportional-band resistor is divided by the setting of the proportional-band adjustment, and the result appears as a negative-feedback signal to the forward-amplifier input. For example, without reset and derivative, an input step-change appears in the feedback loop and is stepped up 4 to 1 by the feedback transformer. If the proportional band is set at 400 percent, all the feedback appears at the amplifier input. If the proportional band is set at 2 percent, a minimum amount of feedback appears at the amplifier.

When a proportional-plus-reset controller is used, if the proportional band is set at 100 percent and the reset at 0.5 repeat per min and a step-change of 25 percent of the measured variable is introduced, the output pointer should instantaneously move from the midpoint or 12-ma point to either the 8- or 16-ma value due to the proportional action. Movement of the output pointer will continue toward the 4- or 20-ma end points, due to the reset action, and will reach the end point in 2 min.

If the same settings are made with a proportional-plus-reset-plus-derivative controller, the action of the output meter would be the same; however, the initial change in the input would be 20 times the step-change called for by the proportional-band setting. Derivative decay would be proportional to the derivative setting, to a value called for by the proportional-band setting, and then continued change in output proportional to the reset setting.

The Bristol Company, Division, American Chain & Cable Company, Inc.

PRINCIPLE OF DESIGN: The Bristol Dynalectric controller shown in Fig. 12-10 is a compact, three-mode controller which uses all solid-state circuitry. Its compactness lends itself to fit into the subtray of the Bristol 760 series Dynamaster to provide a completely integrated recorder-controller package shown in Fig. 12-10. It has also been designed to fit

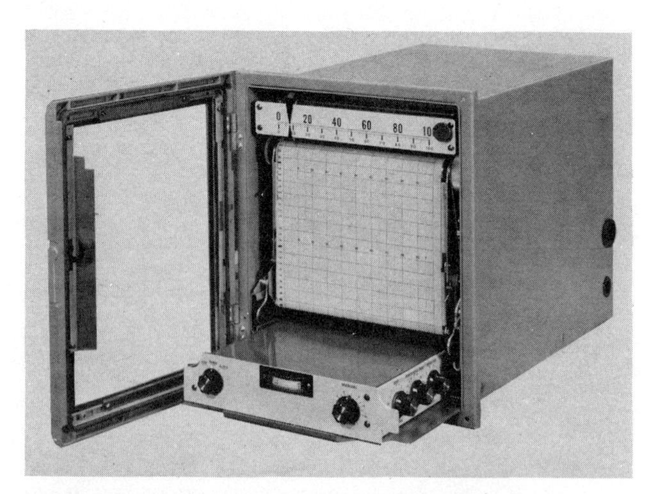

Fig. 12-10 Electronic controller. (*The Bristol Company, Division of American Chain & Cable Company.*)

into the 570 line of Dynamasters or as a separate panel or rack-mounted controller for existing Bristol equipment.

Three models of the Dynalectric controllers are available as described in the following paragraphs. Each model provides a basic type of output suitable for the operation of almost any type of final control element.

Model *J1C*, a current type supplies a 4- to 20-ma dc output to control a magnetic amplifier and saturable reactor, a silicon-controlled rectifier, a current-to-pneumatic converter, or some other current-activated end device. This type of control provides a continuous, stepless, and contactless output specifically adaptable for electrical heating control applications.

Model *J1P*, a position type, has two output relays. The relay contacts are rated for 5 amp at 115 volts ac and are wired to drive a motor-driven valve operator. One relay is used to close the valve, while the other relay drives the valve to an open position. Typical applications would be in the industrial-process market for which an input variable can be controlled by regulating a valve position.

Model *J1D*, a duration type, has as its output a single 5-amp relay contact which is actuated by a time-duration pulse. The pulse varies the relay contact On-time to Off-time ratio as a function of the deviation of the measured variable from set point.

While similar to basic on-off control, duration-type control provides proportional, integral, and derivative actions for a more uniform control of varying process loads. System applications would include processes with on-off control elements, such as contactors or two-position, solenoid-operated valves.

PRINCIPLES OF OPERATION: The forward part of the circuit is common to all three controller models, whereas the feedback circuits and output stages vary with each type of controller. In the current and the duration types, the feedback voltage is obtained from the controller output, whereas in the position type, the feedback voltage is supplied from a slide-wire mounted on the motorized-valve operator.

Normal controller action is initiated when an error signal is detected at the input to the controller. This error signal, when the Dynalectric with a Dynamaster with internal set point is used, is developed between the contact arms of the set-point and retransmitting slide-wires. Both slide-wires are supplied by a dc voltage source within the controller and are connected as a deviation-bridge network. When the measured signal is equal to the set-point value, the two slide-wire contact arms are in the same relative position, and the bridge is balanced. If the measured signal is at some value away from set point, the bridge is unbalanced, and an error signal proportional to the deviation is applied to the controller.

The error signal may also be derived from arrangements other than previously described. In some installations, a remote set-point potentiometer can be used in place of the set-point slide-wire when separation from the recorder or indicator is necessary. Also, the set-point slide-wire may be substituted for by a signal from a remote programmer whenever a series of prearranged operations are required. The Dynalectric control unit can also be used with equipment other than that in the Dynamaster line, providing the error signal is derived from a voltage source.

At the controller input, the error signal is characterized by a special circuit to obtain partial proportional-band and derivative action prior to entering the controller-amplifier sections. The amplifier, which is common to all three types of controllers, has for its first input stage a special diode-modulating circuit consisting of two diodes that act as variable capacitors. A 100-kc carrier signal is fed to the modulator circuit which, under open-loop conditions, is biased so that a zero dc signal at the input will produce a 12-ma output at the final control element or a 50 percent on-to-off time (with the *J1P* controller) or a balance or no-output condition (with the *J1P* controller). If a dc voltage (error signal) is impressed on the input, the capacitance effect of one diode will increase, while that of the other will simultaneously decrease. This capacitance change causes an impedance unbalance, modulating the carrier in phase and amplitude proportional to the input signal to change the steady-state output of the controller. A portion of the output signal is fed back to the modulator circuits in the case of the *J1C* and *J1D* controllers. The degenerative-feedback loop contains a segment of the proportional-band adjustment and the integral network to complete the three-mode action of the controller.

All Dynalectric models contain the following controls and switches:

Off-Man-Auto transfer selector switch: In the Off position, no ac power is supplied to the controller. In the Man position, instrument output is controlled by manual knob 3 and read on output meter 2. The Auto position transfers the instrument to automatic control, which is governed by the setting of mode controls 4, 5, and 6.

The output meter on the face of the controller, calibrated from 0 to 100 percent, gives information pertinent to each controller:

J1C indicates the current output.

J1D indicates the percentage of time that output contacts of load relay will be closed during a duration-time cycle.

J1P indicates in percentage the position of the final control element as 0 percent for fully closed and 100 percent for fully open.

The Man knob scale is graduated from 0 to 10 which in the case of the *J1C* controls the current (4- to 20-ma) output; *J1D* controls the

On time of the duration pulse from fully Off to fully On; *J1P* controls the position of the final element from fully open to fully closed.

The integral knob scale is graduated from 0 to 100 repeats per min to counteract "droop" or "offset" caused by proportional action.

The proportional-band-knob scale is graduated from 2 to 500 percent and is used to regulate the amount of initial corrective action taken by the controller in relation to the amount of deviation from set point.

The derivative knob is graduated from 0 to 10 min and is used to "speed up" the effects of the proportional action.

Load adjust 7 (*J1C* only) provides a means of matching the external load to the controller output.

Sensitivity (*J1P* only) controls the deadband or sensitivity of the system between the open and close outputs. (Not shown.)

The cycle time knob (*J1D* only) is calibrated from 2 to 150 sec, which is in reference to the duration cycle. (Not shown.)

Fischer & Porter Company

PRINCIPLE OF DESIGN: Figure 12-11 is a Fischer & Porter electronic Scan-Line* recorder which can be single- or group-mounted in a 3-

* Registered trade name, Fischer & Porter Company.

Fig. 12-11 Scan-Line electronic recorder controller. (*Fischer & Porter Company*)

by 6-in. cutout. The recorder accepts electrical analog signals representing variables; the input is recorded on a 4-in. strip chart and displayed on a vertical indicating scale. The instrument may be furnished with facilities for a single- or dual-pen system, each value being exhibited on a separate scale and recorded in different color ink. The input signal is usually in the 1- to 5-volt dc range derived from a current source; however, other ranges may be specified.

The instrument employs a unique driving device called a Torq-Er* which includes a contactless feedback device called a "flux bridge." Both the Torq-Er and associated electronics are modular in design and are easily disconnected from the recorder chassis. In addition, an internal power supply provides operating power for two-wire transmitters.

The process-variable signal is applied as input to a differential amplifier whose output energizes proportional solid-state "up" or "down" gates. These gates in turn apply current to a Torq-Er which positions the pen on the chart. The Torq-Er connects to a flux bridge and velocity generator to provide pen position and to control feedback signal which becomes the second input to the differential amplifier.

The Torq-Er functionally is the prime mover for the recording pen. It consists of a permanent magnet armature that is actuated by the application of a direct current (120 ma maximum) through either an upscale or downscale winding. When balanced, a holding current of 30 ma flows through the up-and-down scale windings. When pen movement is necessary, the current increases in the proper winding to a maximum of 120 ma (if necessary). Rotation of the armature is limited to 35°, equivalent to the minimum-maximum vertical drive of the pen.

The flux bridge is a contactless, solid-state, position-feedback device used in conjunction with the Torq-Er. The primary coils are energized from a 10-kc oscillator; the voltage developed in the secondary coils is proportional to the position of the gates that are directly coupled to the Torq-Er armature shaft. This provides a feedback signal proportional to armature position. A permanent magnet attached to the flux gates operates in conjunction with a velocity coil to provide a rate (proportional to speed) feedback signal. This is applied as degenerative feedback to the amplifier input to reduce the pen speed.

The servo connecting link positions the recording-pen drive arm. A pin at the front of the drive arm engages a slot in the recording-pen carrier. Since the carrier is guided by two vertical rods, the pin moves back and forth as the pen moves vertically. The recording pen can be removed from the carrier for servicing and substitution of another pen.

* Registered trade name, Fischer & Porter Company.

The recorder has a capillary-type inking system which, in conjunction with a special ink formula, provides: (1) a uniform ink line on the recording chart, (2) an instrument that writes with the chart at any angle between vertical and a 75° tilt back, and (3) the convenience of a slip-in ink capsule that carries a large ink supply.

The recording pen connects by way of plastic capillary tubing to the ink capsule. The ink, in a vapor-sealed system, will not evaporate, thereby causing the pen or tubing to clog. The ink capsule fits into a slot in the bottom of the chassis where a space is provided for two capsules. The factory-sealed ink capsule is made of a pliable plastic material. The capsule holder has concentric pierce and prime tubes; the longer pierce tube connects to the recording pen and is used as the ink-feed tube. The shorter prime tube is attached to the vented squeeze bulb that is used to vent the cartridge or to apply pressure to the cartridge. Venting is accomplished upon installation of the capsule. Pressure is applied by closing the squeeze-bulb vent (with a finger) and gently pressing the bulb. Applying pressure to the capsule forces ink to the recording pen.

The chart-drive motor operates the timing drum and storage roll by way of timing belts beneath the chassis. The paper leaving the chart feed roll passes around the plate and timing drum. Sprocket teeth on the timing drum engage sprocket holes in the chart paper to assure that the paper passes the recording pen at a uniform rate of speed. Uniform chart tension is provided by three fingers that project from a bar on the pen assembly.

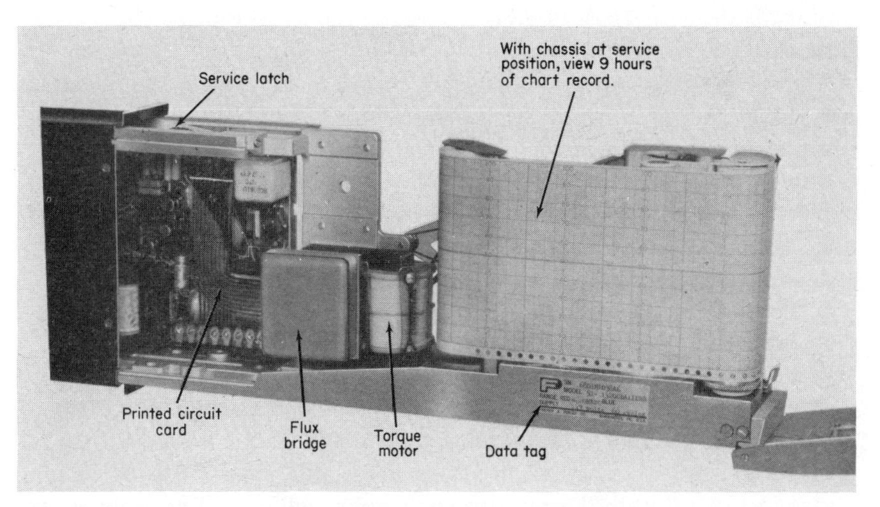

Fig. 12-12 Chassis partially withdrawn from instrument case. (*Fischer & Porter Company*)

The normal chart speed is ⅞ in. per hr. When the chassis is drawn to the service position, approximately 9 hr of the previous recording may be viewed. The timing drum may be moved forward by hand to set the chart for the correct time. The drum cannot be rotated in reverse.

The chart-drive motor and electronics are protected by a ½-amp fuse. Recorder models operating from a 24-volt dc source have two fuses: one ½ amp for the 24 volts dc to the electronics, and one ½ amp for the chart motor power. The fuses are labeled dc and ac, respectively, on the rear terminal board. An On-Off chart-drive power switch is also available as an option. The dual-speed chart motor option has a three-position switch for selection of slow, off, or fast speeds.

The recorder chassis may be partially withdrawn from the instrument case to a locked service position, as shown in Fig. 12-12, or completely removed for servicing or substitution of another chassis.

Description of Electronic Control Instruments

The Foxboro Company

Principle of design: Foxboro electronic Consotrol controllers exemplify modern panel control equipment, with features particularly adaptable to present-day practice. The fully transistorized instruments have no vacuum tubes and no moving parts, except indicators and operator adjustments. They provide feed-forward and other advanced control procedures. They are fully compatible with digital-computer process control, both computer-supervised operation and direct-digital-computer control. Safe and dependable electronic Consotrol controllers are available with Underwriters Laboratory, Inc., listing as elements of intrinsically safe control loops.

The series 62H Consotrol controller, shown in Fig. 12-13, is a universal indicating control instrument; the series 61H controller is a simplified unit specifically designed for economical flow control.

The Foxboro series 62H universal controller: The series 62H controller is available in a number of variations for general and special purposes. This controller is available for:

1. Simple proportional control
2. Proportional-plus-integral (reset) control
3. Proportional-plus-derivative control
4. Proportional-plus-integral (reset) -plus-derivative control

Proportional action is adjustable from 300 to 5 percent proportional band (0.3 to 20 gain). Integral action is adjustable from 0.015 to 30 min per repeat. Derivative action is adjustable from 0.015 to 30 min and is incorporated in the measurement input, which avoids exaggerated changes of controller output resulting from change of set point. These adjustments are available on the mode panel, shown in Fig. 12-13, on the side of the controller, accessible from the front of the panel by pulling the controller forward without disturbing its operation.

Series 62H controllers can perform:

1. Single closed-loop control.
2. Single unit-ratio control; this can be switched to direct single-loop control by a switch on the mode panel. The ratio dial becomes the set-point dial in single-loop control.
3. Cascade control, with a switch for transfer from remote set point to controller set point.
4. Autoselector or auctioneering control, with automatic or manual selection of controlling variable.
5. High-limit, low-limit, or high-low-limit control. Available with built-in power supply for operation of electronic- or differential-pressure transmitters.
6. Batch operation for automatic start-up without overshoot of control point. Separate adjustment to provide fastest possible start-up and optimum steady-state operation.
7. Digital-computer-supervised control, with digital input to set-point adjustment.
8. Direct-digital-computer-control backup, with remotely operated, self-balancing transfer from computer- to local-controlled operation.

OPERATION—SERIES 62H UNIVERSAL CONTROLLERS: *Measurement signal.* The block diagram shown in Fig. 12-14 shows principal elements of the series 62H controller. The measurement signal is a 10- to 50-ma

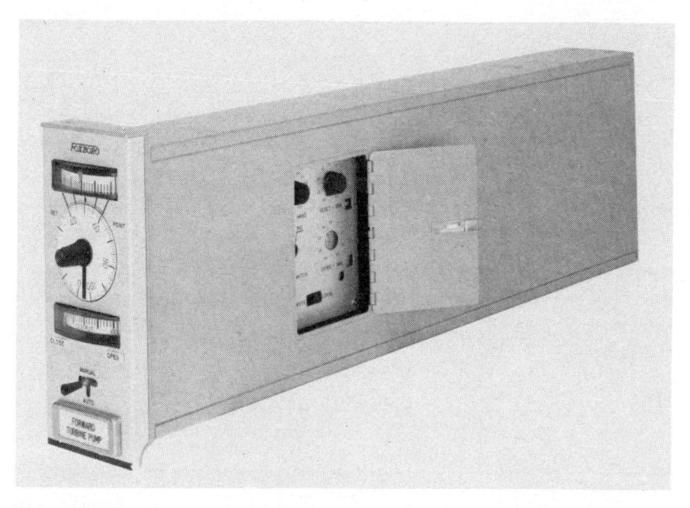

Fig. 12-13 Consotrol 62H controller. (*The Foxboro Company*)

dc current; this current flows through a 100-ohm resistor which forms a part of the input bridge, developing a 1- to 5-volt dc potential.

Derivative action. The 1- to 5-volt dc potential is also applied to the input of the separate (optional) derivative amplifier. The dc output from the derivative amplifier flows through another fixed resistor in the input bridge, developing a dc potential proportional to the rate of change of measurement. In many earlier controllers, derivative action was developed in the feedback loop of the controller. The series 62H circuit avoids the objectionable set-point reactions of the earlier types where an exaggerated change of output resulted from set-point changes.

Set-point signal. A set-point signal is developed by a 2- to 10-ma current flow through a 500-ohm resistor in series with the 100-ohm measurement resistor in the input-point bridge. Thus current develops a 1- to 5-volt potential of opposite polarity to the measurement potential. The current is derived from a double-diode-regulated source and a set-point potentiometer controlling a set-point transistor. A 100-ohm resistor in this set-point circuit provides for a 0.2- or 1-volt signal for remote readout of set point.

Set point may also be developed by a 10- to 50-ma dc signal from another controller, a separate set-point station, etc. For operation with digital computers, a motor drive for the set-point potentiometer is available.

Proportional action. The input bridge consists of fixed resistors for measurement, derivative, and set-point signals and additional fixed-balancing resistors. Feedback potential from controller output is applied

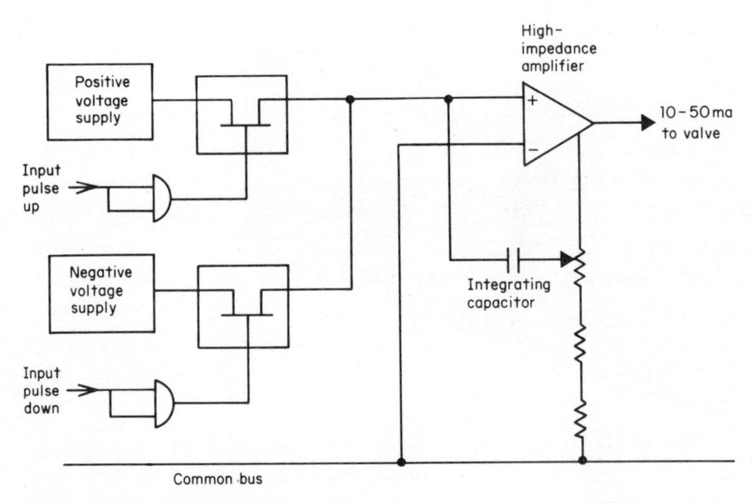

Fig. 12-14 Block diagram of 62H electronic controller. (*The Foxboro Company*)

across opposite corners of the bridge. This feedback potential is adjustable, providing adjustment of proportional action. This is a voltage-balancing rather than a resistance-balancing bridge; all resistances in the arms of the bridge are fixed.

Integral (reset) action. Integral (reset) action is developed by a fixed capacitor and variable resistance, connected in the output circuit of the bridge. Ten-mfd capacitance and 200-megohm resistance (maximum) gives up to 2,000-sec, time constant, 0.03-repeat-per-min integral control action. The controller action, through the feedback circuit operation and bridge circuit, maintains substantial balance, with a small (1 mv or less), residual, unbalance output potential from the bridge throughout the operating range.

Positive feedback. The Varactors in the oscillator bridge change their capacitance with change of applied dc potential. These Varactors form two arms of the oscillator bridge circuit; change in capacitance alters the unbalance of this bridge. This bridge forms part of a positive-feedback circuit connected around the high-gain ac amplifier. Bridge output is fed through a 130-kHz resonant circuit to amplifier input, inducing an oscillation whose amplitude depends upon the positive feedback. Change in the residual dc unbalance potential from the input bridge applied to the Varactors alters their capacitance and thereby the unbalance of the oscillator bridge. This controls positive feedback and the ac amplifier output. A change of dc potential of approximately 1 mv is sufficient to change output from minimum to maximum normal amplitude.

Manual control. The ac and dc amplifiers also generate the output signal for manual-control operation. In the Manual position, the transfer switch disconnects the oscillator bridge from the input bridge and, instead, connects it to the negative-feedback potential from the output circuit and to a manual-setting capacitor. The bridge, amplifiers, and output circuit maintain feedback potential substantially equal to capacitor potential; the dc potential applied to the oscillator bridge is maintained at less than 1 mv. Thus, output current is precisely established by the potential of the manual-setting capacitor. The previously mentioned, very high, effective input resistance of the oscillator bridge combines with a high-quality manual-setting capacitor so that this capacitor retains its charge over a long period, with a drift rate corresponding to less than 1 percent per hr change in controller output under manual-control operation.

In manual operation, output is raised (or lowered) by increasing (or decreasing) the charge on the manual-setting capacitor and hence its potential. A small current flow from a voltage source and high resistance increases (or decreases) charge at such a rate that output changes at 1 percent full scale per sec. A second switch position provides a larger current for full-scale output change in 6 sec.

The transfer switch. The transfer switch, shown in Fig. 12-13, is a tee-bar toggle structure which transfers between automatic and manual operation and also adjusts output when in manual operation. Setting the toggle down to Auto places the controller in automatic operation. Setting the toggle up to Manual places the controller in manual operation. In the Manual position, the toggle can be moved horizontally against a spring return, which restores the switch to center position when released. There are two positions on each side of center, with the intermediate identified by a readily recognizable spring detent. In this intermediate position, manual-control output changes by 1 percent per sec. In the extreme position, output changes by 17 percent per sec. The circuit for this was described in the previous paragraph.

Actual value of output is displayed both in automatic and in manual operation by an indicator located immediately above the transfer switch. This arrangement provides a simple and convenient transfer and adjustment under manual control.

For operation with direct-digital computer control, a "cross" instead of a tee-bar toggle is used. The added Up position is for computer control. In this position, the charge on the manual-setting capacitor is controlled by a digital signal from the computer. This forms a convenient link between the digital-computer signal and the analog signal required for conventional valve position, etc. Manual and automatic operation are as previously described. In case of computer failure, transfer from Digital position to Manual or Analog Automatic position can be done by a relay within the controller.

THE FOXBORO SERIES 61H FLOW CONTROLLER: The series 61H controllers are simplified controllers, designed specifically for flow control and similar rapid-response control operations. They are available only with proportion-plus-integral (reset) action, with a proportional band of 100 to 300 percent (gain of 1.0 to 0.3). For this flow control function, balanceless transfer between automatic and manual control is omitted, since the flow rate reaches its final value in a matter of seconds in both automatic and manual modes. For the same reason, "batch" operation is not included. These controllers are available for cascade operation. A power supply for operation of a pressure or differential-pressure transmitter is standard in the series 61H controllers.

The series H controller includes a number of auxiliary stations, such as the controller bypass station previously described. Separate manual-control stations for both set point and output, ratio stations, etc., provide the flexibility of operation demanded by modern, sophisticated process control.

OPERATION—SERIES 61H FLOW CONTROLLER: Operation of the series 61H flow controller, shown in Fig. 12-15, is basically similar to the series 62H universal instrument. Output, both in automatic and manual operation, is provided by a transistorized dc output power stage controlled by an operational amplifier. A differential dc amplifier algebraically adds measurement set-point (or manual-set) signal to feedback signal; the small differential-unbalance output is connected to the input of the operational amplifier. Feedback is direct in manual operation; in automatic control operation, a capacitance and two variable resistors in the feedback circuit permit adjustable proportional and integral (reset) action. All dc amplification is used. This, with the low-impedance integrating circuit permitted by the 30-sec maximum time constant and the fact that maximum required controller gain is 1, makes possible a simple electronic circuit, without any sacrifice of the performance required for flow control and similar applications.

As a consequence of all dc amplification, input and output circuits of the series 61H controller are not dc-isolated. However, these controllers are almost always applied to pressure and differential-pressure control; an individual power supply for the pressure or differential-pressure transmitter is standard in all series 61H controllers. This renders isolation of a measurement input circuit unimportant. In this controller, the set-point resistor is directly connected to the common connection line of the instrument (essentially a floating "instrument ground"). When the controller is used as the secondary in a cascade control system, with the output of the primary controller grounded, this connection establishes a suitable common ground for both instruments.

Fig. 12-15 High-speed electronic controller. (*The Foxboro Company*)

Honeywell, Incorporated

PRINCIPLE OF DESIGN: Figure 12-16 is a supervisory control of the modular construction designed by Honeywell, Incorporated, which is referred to as "VSI for supervisory control." It includes a variety of optional modules such as a 42-volt dc power supply; low-alarm, process-variable monitor; high-alarm, process-variable monitor; deviation alarm monitor; and reset limiter which can be readily installed or removed. The controller is easily interchangeable with others of its type to provide one-, two-, or three-mode control.

The VSI for supervisory control consists of a case, a subtray, and chassis assemblies. The subtray and chassis assemblies slide forward to a service position where they lock into place. They can be completely removed from the case by depressing a release latch. The subtray and chassis assemblies are held together by two large knurled screws on the underside of the subtray. When these screws are loosened, the chassis assembly can be separated from the subtray. The VSI for computer control can be operated in the manual mode with the chassis assembly removed, but there will be no output indication. The manual loader module permits the chassis assembly to be replaced while manual control of the process is maintained.

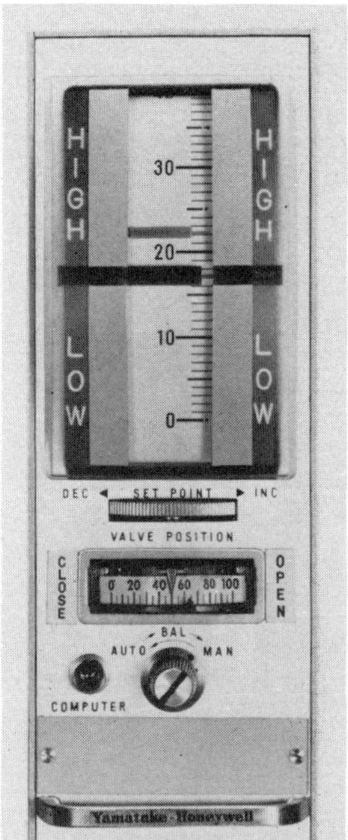

In the computer mode of operation, Loc/Bal/Comp switch S1 is set to position *C* (computer) and the relay *K*5 is energized. The contacts of *K*5 are in the state shown in Fig. 12-17. External set-point signal I_{ES} and internal set-point signal I_{IS} are applied to the servo amplifier. Whenever the value of I_{ES} is different from I_{IS}, it is detected by the servo amplifier. The output of the servo amplifier drives the servomotor which, in turn, adjusts the internal set-point generator until I_{IS} equals I_{ES}. Thus, the internally generated set point always follows the external set point received from the computer.

Process-variable signal I_{pv} and I_{ES} are applied to opposite ends of the deviation meter-resistor *R*3 circuit. The deviation meter provides a visual indication of the set point. If there is any difference between I_{pv} and I_{ES}, a differential current flows in *R*3 generating a voltage drop across *R*3. The signal developed across

Fig. 12-16 Supervisory control station. (*Honeywell, Incorporated*)

*R*3 is the error-signal input to the controller. The controller-output (4- to 20-ma dc) signal adjusts the final control element until the process variable equals the computer set point (external set point).

While the computer is functioning normally, the computer-failure contact, located in the computer, is closed and relay *K*5 is energized. All contacts of the relay are in position *C* (computer). Contact *K5a* connects the set-point signals to the servo-amplifier input. Contact *K5b* lights the Computer indicator on the front panel of the VSI, informing the operator that the process set point is being supplied by the computer. Contact *K5c* provides a dry contact input to the computer for use by the computer programmer. Contact *K5d* is a holding contact for relay *K*5. Internal set-point signal I_{IS} is fed back to the computer. This signal can be used by the computer programmer to make certain that I_{IS} is tracking I_{ES}.

Whenever there is a computer failure, the computer-failure contact opens and relay *K*5 deenergizes. The switching of contact *K5a* to the L (local) position disconnects the internal and external set-point signals

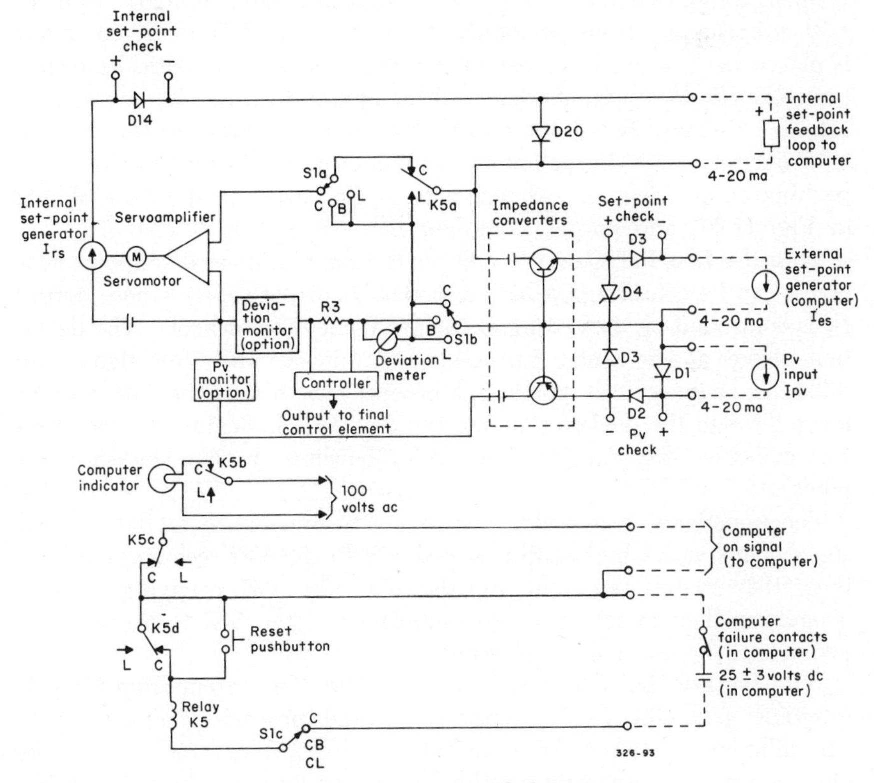

Fig. 12-17 VSI for supervisory control, simplified schematic diagram. (*Honeywell, Incorporated*)

from the servo-amplifier input. The external set-point signal no longer flows in the VSI because its current path is broken by the open *C* contact of *K5a*. The internal set-point signal is applied to *R*3 through the now closed L contact of *K5a*. The error signal developed across *K*3 is now being generated by the internal set-point signal and the process-variable signal. The internal set-point signal is at the value of the external set-point signal just before computer failure. Essentially, the VSI is in the local control mode of operation and establishes the set point for the process. Since the internal set-point generator has been tracking the external set-point generator, the transfer from computer to local mode of operation is bumpless. The foregoing circuit action prevents a false output from the computer from upsetting process control.

The Computer indicator on the VSI extinguishes when contact *K5b* switches to position L. The extinguishing of this indicator informs the operator that the computer is malfunctioning and that the VSI is establishing the process set point. The switching of contact *K5c* to position L opens the dry-contact Computer On signal to the computer. Contact *K5d* also changes from position C to position L. When the computer is placed back on line, the computer-failure contact automatically closes, but relay *K*5 does not energize until the operator presses Reset pushbutton S3. Pressing the Reset pushbutton energizes *K*5, and the relay is held energized by the action of holding contact *K5d* when the Reset pushbutton is released. All contacts of *K*5 revert to the state shown in Fig. 12-17, and normal computer mode operation is resumed.

With the Loc/Bal/Comp switch set to Loc, the internal set-point generator can be considered a high-impedance current source whose output I_{IS} is controlled by the setting of the Set Point thumbwheel. The deviation meter, a zero-center galvanometer, indicates the error signal, the difference between set point and process variable. When no current flows through the deviation meter, the deviation indicator (red pointer) will coincide with the Set Point index hairline (hairline indicates set point).

The impedance converters present a low impedance to the external signal loops and a high-output impedance to the VSI receiving circuits; thus, they match the high impedance of the VSI receiving circuits. Therefore, they match the high impedance of the VSI to the low impedance of the external signal loops.

With the Loc/Bal/Comp switch set to Loc, the current from internal set-point generator I_{IS} is compared to the process-variable current. Any difference between these currents is displayed as a process-variable deviation from the set point on the deviation indicator. In this position, the output of the controller is developed by the difference between I_{IS}

and I_{pv}. The servo amplifier and servomotor are cut off. The magnitude of I_{IS} can be adjusted by the Set Point thumbwheel.

With the Loc/Bal/Comp switch set to Bal, the output of internal set-point generator I_{IS} is compared with the output of external set-point generator I_{ES}. The deviation meter is used as the balance indicator and the output of internal set-point generator I_{IS} is made equal to the output of external set-point generator I_{ES}. The input to the controller is the difference between the internal set point and the process variable, even though the Loc/Bal/Comp switch is in the Bal position. Thus control is maintained in local automatic when the Auto/Bal/Man switch is in the Auto position.

With the Loc/Bal/Comp switch set to Comp, the output of external set-point generator I_{ES} establishes the set point for the VSI. The operation of the VSI is identical to that described in the first part of this section. I_{ES} is continuously compared with I_{IS} at the input of a servo amplifier. If there is any difference between these signals, the output of the servo amplifier adjusts the internal set-point generator output so that I_{IS} follows I_{ES}. The servomotor drives the scale to indicate set point while the internal set point is being equalized with the external set point.

OPERATION IN AUTOMATIC AND MANUAL MODE: Figure 12-17 schematically illustrates controller operation in the automatic and manual modes. The Valve Position meter can be switched into any one of three circuits for manual, automatic, or balance operation. When the VSI is in the automatic mode, the controller output is applied to the final control element. Under manual operation, the manual loader provides the control signal to the final control element.

Automatic mode. When the controller is in the automatic mode of operation, its output of the controller passes through the load (final control element) and the Valve Position meter. The manual loader output is shorted through a relay and flows through $R26$ which provides a load for the manual loader.

Balance mode. This mode of operation provides a means of adjusting the manual loader output when the controller is being switched from automatic to manual operation, thereby allowing a smooth transfer. When the Auto/Bal/Man switch is set to Bal, the Valve Position meter is relocated in the circuit. This meter will now indicate the current difference between the controller output and the manual loader output. The controller output is maintained during the balancing operation, and the output provided to the load does not change. Through adjustment of the Manual thumbwheel, the output of the manual loader is made to equal the controller output, thereby permitting the transfer.

Manual mode. In this mode, the final control element is actuated by the manual loader output. This output is displayed on the Valve Position meter and can be adjusted by the Manual thumbwheel. Resistor *R*26 shunts the controller output and acts as a load for the controller. Since both the controller and the manual loader output currents flow through *R*26, the net current flowing through it is the difference between the output current levels. Resistor *R*26 provides a feedback to the controller, which causes the controller output to change until the voltage drop across feedback resistor *R*26 is 0 volts. This action permits a shift from manual to automatic modes without the controller's having to go through a balancing operation.

In-Val-Co, Division of Combustion Engineering, Inc.

PRINCIPLE OF DESIGN: Figure 12-18 is a Gamm-O-Switch of the In-Val-Co design. The switch is completely self-contained and requires no interconnecting remote amplifier or power unit. The housing is small, lightweight, and fully explosion-proof with an easily removable, screw-type cover (see Fig. 12-19). The chassis is a plug-in module and makes use of the latest advances in printed and transistorized circuitry. There is only one simple adjustment for ease in calibration and

Fig. 12-18 Gamma-ray switch. (*In-Val-Co, Division of Combustion Engineering, Inc.*)

start-up. In-Val-Co's proved radiation counter tube is used in conjunction with temperature-stable silicone transistors to offer maximum reliability under the most severe operating conditions.

The Gamm-O-Switch is available in two models. Model 200 LLFS is low level and fail-safe (relay deenergized with no level) and model 200 HLFS is high level and fail-safe (relay energized with no level).

A source holder, containing a source such as radium 226, cobalt 60, etc., is normally positioned on one side of a vessel or pipe and the Gamm-O-Switch on the opposite side. Thus with no material in the vessel, maximum radiation is received by the Gamm-O-Switch. When the vessel fills to the control point, the gamma rays are absorbed by the material, which in turn causes the relay in the Gamm-O-Switch to operate. Thus the Gamm-O-Switch may be used for high- or low-level indication or control. Interface level between two materials may also be detected, provided there is sufficient specific-gravity difference.

Nuclear-level controls are particularly advantageous as most applications permit both the source and holder to be mounted externally to the vessel and therefore are unaffected by corrosion, abrasion, pressure, and sanitation requirements. This also eliminates the need for shutdown to service or install the instrument. The Gamm-O-Switch detects

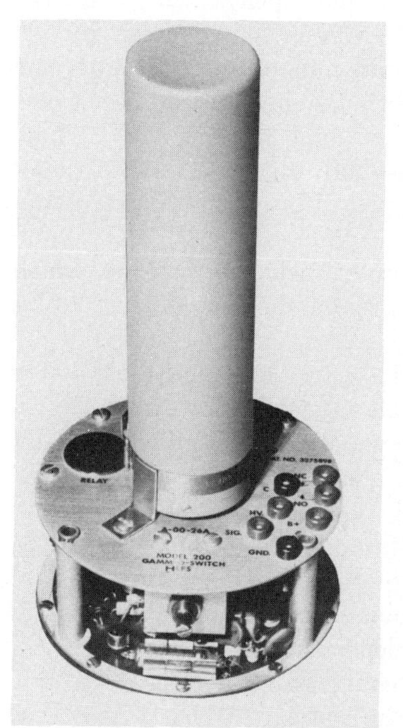

Fig. 12-19 Gamma-ray switch with cover removed. (*In-Val-Co, Division of Combustion Engineering, Inc.*)

virtually any liquid, solid, or slurry. The correct radioactive-source size must be determined for each specific application. Upon receipt of full installation details, In-Val-Co's experienced application engineers will gladly submit recommendations for any particular problem. This information should include vessel or pipe-wall thickness and diameter plus the name and specific gravity of the material to be detected.

Radium 226 and cobalt 60 are the most commonly used radioactive-source material. Radium is normally used for applications requiring 10 mg or less. Radium has a half-life of 1,620 years and does not require an AEC license for its use. The half-life of a radioactive material is the time required for its output in gamma rays to diminish to one-half its original value. Due to the very high sensitivity of the Gamm-O-Switch, extremely small sources can be used, ranging from 0.1 to 10 mg of radium. For applications requiring more than 10 mg, cobalt 60 is preferred since it is less expensive for these larger sizes. Cobalt 60 has a half-life of 5.26 years and requires that the user have an AEC by-product material license for its use. In-Val-Co will gladly supply partially completed forms for the application of such a license.

In-Val-Co, Division of Combustion Engineering, Inc.

PRINCIPLE OF DESIGN: There are two principles upon which gamma-ray level-measuring instruments are designed. One method converts gamma rays emitted by some radioactive material into electrical energy; the other method uses a geiger-counter tube where an electrical-output current is proportional to the gamma rays to which it is exposed. These two methods have one thing in common—both use some form of radioactive material for their gamma-ray supply. The most common materials used to supply the gamma rays are radium, cobalt 60, and cesium 137. These materials are selected because of their long half-life, which varies from 5 to 1,585 years, depending upon the material used plus availability.

Figure 12-20 is a circuit diagram of the geiger tube and preamplifier for a geiger-tube-type instrument, which is standard with In-Val-Co and others. Of course, each manufacturer adapts the principle to his particular instrument, but as long as a geiger tube is used, the basic principle will remain the same. In the geiger-tube-type instrument, gamma rays cause electrical discharges through the counter tube. The number of these discharges per min is proportional to the intensity of the radiation penetrating the counter tube. The number of discharges or counts detected by the counter tube depends upon the size of source, the distance between the source and the counter tube, and the absorption of the material between the counter and the source.

Leeds & Northrup Company

PRINCIPLE OF DESIGN: Shown in Fig. 12-21 is a Leeds & Northrup Speedomax M Mark II recorder.* This instrument is a null-balancing, servo-potentiometer type of miniature instrument offered by Leeds & Northrup Company to continually indicate and record the value of a measured variable. Equipped with a 4-in. calibrated scale providing horizontal indication, it is available for one-, two-, or three-channel recording and can be used in conjunction with a variety of transmitters having compatible current or voltage outputs, with spans from 0 to 200 microamperes, up to 0 to 50 ma, or from 1 to 200 volts.

The 4-in. calibrated strip chart is rectilinear and vertically driven at a standard chart speed of ½, 1, 2, 3, or 6 in. per hr. It feeds from a 62-ft supply roll, sufficient for 31 days of operation at a chart speed of 1 in. per hr. Distance to the end of the chart is continuously indicated on each roll for the last 6 ft (three days) of operation. One of several faster speeds is optionally available as a second speed on a two-speed chart drive.

As an option, the Speedomax M Mark II is offered with one or two low-level preamplifiers, adding millivolt capability to the recorder.

* Registered trade name, Leeds & Northrup Company.

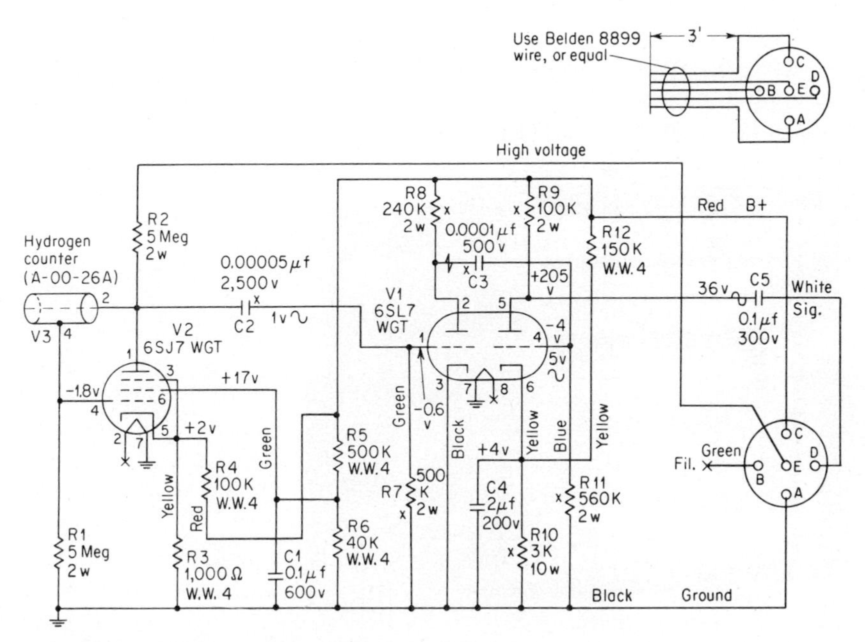

Fig. 12-20 Circuit diagram of geiger tube and preamplifier. (*In-Val-Co, Division of Combustion Engineering, Inc.*)

With this option, the Mark II will accept temperature-compensated thermocouple inputs or direct millivolt inputs with spans from 3 to 66 mv.

The preamplifier(s) bring added flexibility to the recorder, permitting combinations of both high-level and millivolt channels in one compact instrument. On a three-pen model, one high-level and two millivolt channels (or vice versa) can be provided; on a two-pen model, one high-level and one millivolt channel, or two millivolt channels; on a one-pen model, either, as specified.

If control of the measured variable is also desired, the preamplifier(s) can be supplied to provide a 0- to 4-volt-output control signal, compatible with Leeds & Northrup model C control instruments. Removing power from the recorder does not interrupt the control signal, permitting routine maintenance on the recorder without disturbing the control of the process.

Mark II design features have been developed to provide many useful benefits and advantages.

1. Application versatility is one of the outstanding features of this recorder, a wide choice of current- and voltage-input spans gives great freedom in using transmitted current and voltage signals and/or direct millivolt signals, when used with optional preamplifier.

2. As many as three separate circuits can be provided, permitting one-, two-, or three-pen recording of signals from the same or different types of primary elements over the same or different ranges.

3. Zero and span adjustments facilitate its use as a trend recorder. Overlapping records can be separated on a chart to display distinct trends of two or three different variables.

Fig. 12-21 Electronic recorder. *(Leeds & Northrup Company)*

4. Recorder design has the latest state-of-the-art electronic circuits and solid-state components, both of which help assure long in-service dependability and high accuracy.

5. Simplified chart replacement and large ink supply combine to reduce service time and frequency.

6. Combining pen, pointer, and slide-wire contactor into an integral unit provides perfect alignment and assures accurate recording of measured variables.

7. Simple and easy maintenance results from minimized stock of spare parts because units such as the amplifier cards, balancing motors, pen assembly, and reservoirs are interchangeable.

Figure 12-22 is a Mark II partially withdrawn from the case. The paper-drive mechanism can be released from the chassis by a pushbutton. The pushbutton when pressed allows the chart mechanism to be removed for changing or inspecting the chart.

The compact Mark II is especially advantageous in applications where panel space is at a premium and where minimum capital outlay is essential, at no sacrifice of performance. Some of the variables which can be measured and recorded, and controlled if desired, include tempera-

Fig. 12-22 Front view of electronic recorder partially withdrawn from case. (*Leeds & Northrup Company*)

ture, volts, millivolts, current, pressure, flow pH, redox, percentage of oxygen, wind direction and velocity, etc.

Additional flexibility is made possible since the recorder can be provided with one-, two-, or three-pen recording of input signals from the same or different types of primary elements over the same or different ranges.

The various features discussed above list only some of the major features; many other features of less importance are available.

Leeds & Northrup Company

PRINCIPLE OF DESIGN: Figure 12-23 is a solid-state Electromax* signaling controller of the Leeds & Northrup design. This controller gives the accuracy and reliability of a set-point potentiometer combined with a drift-free amplifier and feedback control circuit for thermocouple temperatures up to 3000°F. When the setter knob is pressed and turned, the control point is set to digital readout of 1°F or C. Deviation of the input signal from the set point is clearly displayed on the zero-center meter.

Small and compact, the Electromax design is relatively new on the market. However, it is filling a gap long needed by industry because its cost range is appealing, and simple maintenance makes it suitable for small jobs where cost is a major factor.

The Electromax controller comprises a set-point potentiometer, ac and dc amplifiers, modulating and demodulating converters, and feedback circuits appropriate to the control form. It is supplied as a two-position, a current-adjusting type or a position-adjusting type, controller.

OPERATION: The input to the controller is a millivoltage signal from a thermocouple (or other emf source). This thermocouple emf, for

* Registered trade name, Leeds & Northrup Company.

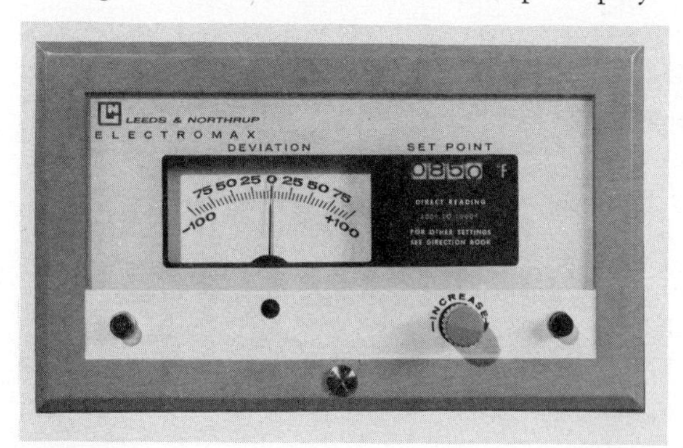

Fig. 12-23 Front view of electronic indicating controller. (*Leeds & Northrup Company*)

example, is opposed by the set-point emf, representing the temperature at which the process is to be controlled.

Set-point potentiometer. The multi-turn set-point potentiometer is geared to a four-digit counter to provide a 1,000-count span (primarily for celsius ranges) or an 1,800-count span (for Fahrenheit and the 18-mv ranges).

The calibrated ranges are of two types:

1. Ranges utilizing linear potentiometers, adjusted so that accuracy of set-point setting is within ±0.5 percent of input span at two calibration points and within ±1 percent of span (for most ranges) between them

2. Ranges involving the use of characterized potentiometers to match the thermocouple curves within ±1 percent of input span over the entire range

Detector circuitry. Any deviation of the thermocouple emf from the set-point emf is filtered and applied to the modulator. The circuit between the modulator and deviation meter functions as a feedback amplifier to produce a feedback voltage that is opposite, but substantially equal, to the deviation voltage. The overall gain of this type of amplifier is low but very stable. The very small difference between the deviation and feedback voltages is converted to ac at line frequency by the modulator, amplified by the differential ac amplifier, converted back to dc by the demodulator, and further amplified by the first differential dc amplifier.

The output of the amplifier is measured by the zero-center deviation meter, graduated −100 to 0 to +100. For any temperature range, the meter reading corresponds to the deviation in degrees. For the millivoltage range, the meter reading divided by 100 represents the deviation in millivolts. If the input emf is less than the set-point emf, the meter reading is to the left of zero center; if the input emf is more than the set-point emf, the meter reading is to the right of center.

Current-adjusting-type control. In the current-adjusting-type controller, the amplified bipolar deviation signal is further amplified by the second differential dc amplifier and applied to the output stage. In addition to flowing through an external magnetic amplifier or other current-receiving device, the output current also develops a signal that is acted upon by the proportional-band and reset circuits and fed back to the input of the second differential dc amplifier where it opposes the deviation signal. In this manner the current is adjusted to, and maintained at, the level that is required to make the input emf equal to the set-point emf. For power failure to the controller, the output current drops to zero to deenergize the final control device.

Position-adjusting-type control. In the position-adjusting-type controller, a portion of the amplified bipolar deviation signal, as determined by the proportional-band setting, is further amplified by the second differential dc amplifier and applied to output relays. The output relays operate a reversible drive-unit motor which positions the control valve. The drive unit also positions a contact on a feedback slide-wire. This develops a signal that acts upon the reset circuit and is fed back to the input of the second differential dc amplifier where it opposes the deviation signal. In this manner the control valve is positioned as required to make the input emf equal to the set-point emf. For power failure to the controller, both relays drop out, opening the "raise" contact and closing the "lower" contact to drive the value to its closed position.

Leeds & Northrup Company

PRINCIPLE OF DESIGN: Figure 12-24 is a Leeds & Northrup Millitemp current-adjusting-type controller which is a thermocouple-actuated millivoltmeter. Providing reset and proportional-control actions, the Millitemp controller indicates process temperature (up to 3000°F) and controls it by continuously regulating power input to the heating element. Reset rate is fixed, 0.05 repeat per min, and the proportional band is adjustable over a range of 0.2 to 3.6 mv (approximately 140°F for current-adjusting couples). This controller produces a 0- to 5-ma output which is fed to a saturable-core reactor or silicon-controlled rectifier power package to regulate power input to the process.

Of its many features, one of the most outstanding is the taut-band system of suspension. This type of suspension eliminates bearing friction, a major cause of trouble experienced in the use of conventional pivot-and-jewel bearings. In conventional systems, pivots wear and become flat and jewels get damaged and dirty—this is a major cause of

Fig. 12-24 Current-adjusting-type controller. (*Leeds & Northrup Company*)

inaccurate readings. There is no need to tap a Millitemp case to make sure the pointer is not sticking. Another major feature is the inclusion of automatic reset which assures that the controlled temperature will be held at the desired set point regardless of changes in the load.

Solid-state circuit components are also featured, along with automatic reference-junction compensation which compensates for the effects of ambient temperature changes at the controller.

Automatic shutdown is provided should an "open circuit" occur in the sensing-element circuit. Indication is provided by an indicating pointer and a 5-in. scale.

Millitemp instruments have a mirror behind the indicating pointer to remove parallax error, adding to the ease and accuracy of reading the meter. Readability is further enhanced by a fluorescent red pointer and a pale-green mask around the glass-enclosed green scale.

These instruments are suitable for many applications in the food, rubber, ceramics, metals, fabricating and processing, and transportation-equipment industries. Actuated with either base- or noble-metal thermocouples, they will measure temperatures as high as 3000°F.

LEEDS & NORTHRUP COMPANY

PRINCIPLE OF DESIGN: The Leeds & Northrup M-line model C null detector is a transistorized modular unit that compares two dc input signals and operates one of two output relays, depending upon the polarity of the difference between the two signals. When the difference between the signals with normal gain adjustment is less than 20 mv, neither relay is energized. When there is more than 20 mv, one relay or the other is energized. The relay contacts are interlocked to prevent simultaneous closure of both output circuits.

The null detector is designed to provide the following features: (1) improved control performance—fast response is provided by high sensitivity and all-electric operation; (2) ease of maintenance—modular plug-in design facilitates component replacement; and (3) reliability—the solid-state design assures long component life with maximum in-service time.

Figure 12-25 is a front view of a Leeds & Northrup M-line model C bias/auto-manual station with two-position meters and a bias setter. These stations for position-adjusting control are compact, modular operating units designed to be used with the model C null detector to favor or equalize final control element participation in a control loop. They feature plug-in convenience, front-of-panel adjustments, and mounting versatility. They can be mounted into a relay rack or on a panel. Standard models comprise the following front-of-panel adjustments and indicating meters: (1) bias setter, two-position meters, three pushbuttons—

Auto, manual "raise" and manual "lower"; (2) bias setter, position meter, three pushbuttons—Auto, manual "raise" and manual "lower"; (3) bias setter, position meter; and (4) bias setter only.

The M-line model C null detector is used in conjunction with the M-line model C bias/auto-maunal stations in a variety of applications, one of which is referred to as "active equalization." Here the null detector operates a second follower-drive unit in relation to a master device which could be considered a reference-drive unit. As the position of the master-drive unit and its slide-wire contact changes, the input to the null detector becomes unbalanced. The null detector sends either "raise" or "lower" signals to drive the slide-wire on the follower-drive unit to a balanced condition. The bias/auto manual station is used to adjust the desired relationship between the positions of the master and follower slide-wire for bridge-balance condition. This is accomplished by inserting a biasing voltage in series with the detector input to shift the balance point of the follower slide-wire with respect to the master slide-wire position.

Robertshaw Controls Company

PRINCIPLE OF DESIGN: Figure 12-26 is a Robertshaw Controls Company design of deviation-type controllers. The model 323 utilizes a 10-in.

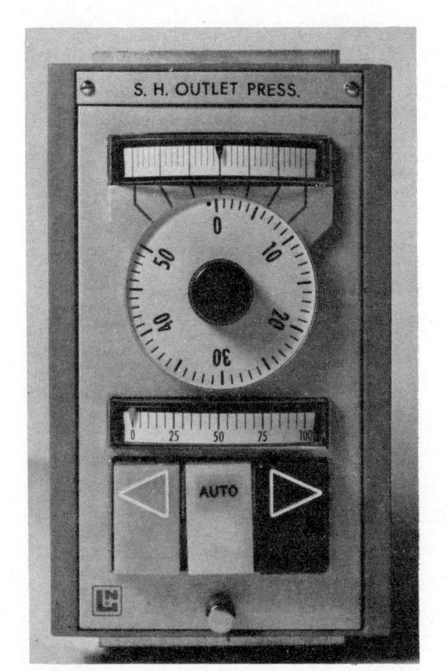

Fig. 12-25 M-Line Bias/Auto-manual station. (*Leeds & Northrup Company*)

Fig. 12-26 Indicating process deviation controller with 10-in. tape scale. (*Robertshaw Controls Company*)

tape-type, indicating, set-point scale and a companion process-deviation indicator adjacent to the set-point scale. Figure 12-27 is a model 324 which provides meter indication of set point with a companion process-deviation indicator. Both controllers are of high-density case design.

The instruments referred to are well constructed, neat in appearance, and simple to maintain. Each unit has its particular use in industry and each receives its input signal from an electronic transmitter which may be used to measure pressure, temperature, differential pressure, liquid level, etc. The instruments are made so that they can be installed singularly or grouped together. They are state-of-the-art designs, replacing the original units which have been on the market for a number of years. The original Robertshaw process control instruments were

Fig. 12-27 Indicating process deviation controller with fixed scale. (*Robertshaw Controls Company*)

known as the American Microsen and were originally distributed by their developer (Manning, Maxwell and Moore, Incorporated). A great deal of research has gone into these instruments to make them a group of reliable process-transmitting and controlling equipment.

The Robertshaw model 323/324 process controllers are solid-state instruments for proportional only, proportional-plus-reset, or proportional-plus-reset-plus-rate control modes. These can be plugged in for easy field change. These controllers offer many features in a process controller such as deviation amplifier, common battery power supply, and new integrated-circuit techniques; "sample and hold" circuitry is optionally available for digital-computer interface. The model 323 employs a long-scale (10-in.) set-point display. The model 324 is a companion controller for cascade inputs using meter display. The controller accepts input signals of 1 to 5, 4 to 20, or 10 to 50 ma dc. The output is available in either 1 to 5 or 4 to 20 ma dc.

A high-impedance-charge (dielectric) amplifier is used in the controller section of the instrument. This permits the use of long-time constants, up to 50 min, in the reset amplifier. A deviation amplifier (located ahead of the control-section, operational amplifier) amplifies the difference between the process-variable and the set-point inputs. This amplified difference (error) signal is applied to the input of the control-section amplifier and is conditioned according to the response setting. The reset and rate-response-circuit network provides feedback to the controller operational amplifier.

The response network, a separate, printed circuit board mounted on the right side of the chassis, contains proportional-band, reset, and rate controls plus an adjustment-hold pushbutton. A cable from this board plugs into the main controller circuit board through a seven-pin, molded plug.

A zener diode is provided across the process-variable-input terminal so that when a controller is removed from an operating loop, input-circuit continuity is not interrupted.

Also a jack is provided on the spectro-strip cable connector (cable between the case and the controller chassis), which enables the operator to provide a continuous current to the valve even when the controller chassis is removed.

The maximum distance of transmission is determined by the sum of the line and the receiver resistance. This total must not exceed 2,500 ohms for 1- to 5-ma output or 700 ohms for 4- to 20-ma output. The open-loop-circuit resistance should be at least 100 times the closed-circuit value to limit the error due to leakage to 1 percent of the transmitter output.

An adjustment-hold pushbutton is located on the response network and should be pressed before any adjustment is made to proportional band or rate. It may also be used for bumpless set-point or input adjustments.

Taylor Instrument Companies

PRINCIPLE OF DESIGN: Figure 12-28 is a front view of a Taylor Quick-Scan* controller which is only 3 by 6 in. in dimension. These instruments can be grouped in almost any number, which can reduce cost installation, panel space, and control-room size. The set point can be set manually or by the computer.

The Taylor series 940R deviation controller is designed for flexibility and accuracy, which meets present-day demand in industry.

Solid-state circuitry is used throughout with components being mounted on a heavy-duty chassis to provide long-term mechanical and electronic reliability. All components, including the motor-driven set point (when specified), are integrated on a 3- by 6-in. instrument slide complete with accordion cable and terminal block suitable for use with any Taylor Quick-Scan housing.

Taylor's proved "human-engineered" features are prevalent throughout the instrument.

1. A green scan-band line provides a ready reference point under which the process set point is established. Any deviation of process

* Registered trade name, Taylor Instrument Companies.

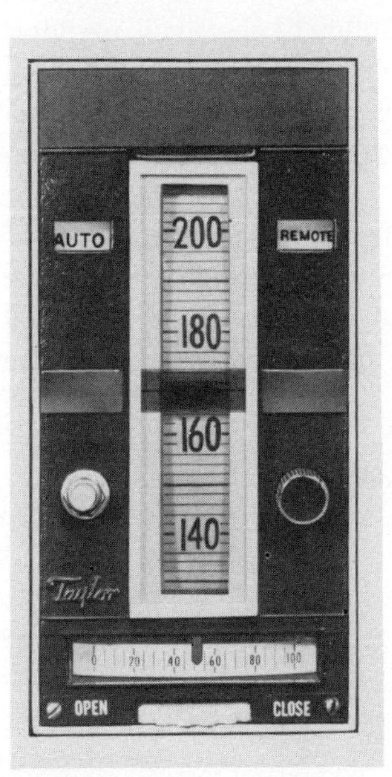

Fig. 12-28 Quick-Scan indicating controller. (*Taylor Instrument Companies*)

from the set point is indicated by a brilliant red pointer which emerges from behind the green scan-band and indicates true-process value. The tape scale is 9 in. long to provide accurate resolution of set-point setting; and 40 percent of the scale is displaced at all times.

2. Two-position, auto-manual, bumpless switching is standard.

3. Two green scan-band alarm lights are optionally available as part of the scan-band feature. Accessories include deviation meter; illumination lights; communications jack; bumpless, four-position, remote-local set-point switching; and transmitter switching.

4. Regulated voltage is supplied to a 10-turn, set-point potentiometer. This front-adjusted set-point potentiometer is synchronized with the 9-in. tape scale. Any error existing between set-point and process voltages is applied to the summing junction at the input of the high-impedance operational amplifier. This amplifier incorporates a field-effect transistor in a bridge circuit to obtain a typical input impedance in excess of 500 megohms.

5. Input-output isolation is standard on all controllers. Power-supply isolation is also provided on ac and isolated dc powered instruments.

6. Pre-Act* (derivative) circuitry is located in the feedback loop for stability together with bumpless gain switching. Output-limiting circuits prevent amplifier saturation and permit start-up without overpeaking on both two- and three-mode controllers. Transmitter power is available at the controller terminal board to supply two Taylor two-wire transmitters. A twisted pair of wires, unshielded, is used to connect the transmitter to the controller and the controller to the final operator, thereby completing a true two-wire control system. Use of optional transmitter switching at the rear terminal block permits power (to transmitter) to be turned off without interrupting the field communication circuits.

A unique feature of the series 940R controller is the provision of a "service-manual" control function on the power supply. The power supply may be detached from the main slide to permit complete removal of the slide for rapid field-servicing without having to shut down the process.

* Registered trade name, Taylor Instrument Companies.

CHAPTER THIRTEEN

Specific-gravity-measuring Elements

In many cases the specific gravity of liquids and gases should be measured and/or recorded. There are instruments on the market today which control the specific gravity of liquids or gases, especially in a blending operation. The art of measuring specific gravity has been perfected, and instruments have been designed (field- or panel-mounted) that are as rugged as other process-control equipment, some of which will be discussed in this chapter.

LIQUID SPECIFIC GRAVITY

Tubular-float-displacement Type of Element

Fisher Governor Company*

PRINCIPLE OF DESIGN: The method of measuring the specific gravity of a liquid by a tubular-float-displacement type of instrument is based on a proved principle of physics (Archimedes' principle) which states that the weight of a body in a liquid is proportional to the liquid dis-

* Grady C. Carroll, *Industrial Process Measuring Instruments,* 1st ed., McGraw-Hill Book Company, New York, 1962, pp. 201–209.

placed by the body times its specific gravity. A familiar instrument based on this principle is the hydrometer; therefore the principle is not new, but only in recent years has it been adapted to industrial-type instruments.

To adapt Archimedes' principle to an industrial-type instrument required a means for submerging a body in a liquid and accurately weighing it with some type of scale so designed that the weight of the body could be transmitted by mechanical means to an industrial-type indicating or recording instrument which could be calibrated to read the specific gravity of the liquid in which it is submerged. A method for doing this has been successfully developed and is now standard equipment with several instrument-manufacturing companies. Since the same principle is used by most of the equipment manufacturers, the major difference being the methods used to accomplish the same end result, only Fisher Governor Company's instrument known by the trade name of Level-Trol* will be used to explain the principle of design.

A unique method for transmitting the apparent change in weight of the float has been developed and is shown in Fig. 13-1. It is known as the "torque-tube method." Since the torque-tube unit is one of the most important items in the instrument, it will be discussed first.

Figure 13-2 shows an assembly view of the torque-tube unit which consists of float rod 1, float-rod driver 2, driver bearing 3, and torque

* Registered trade name, Fisher Governor Company.

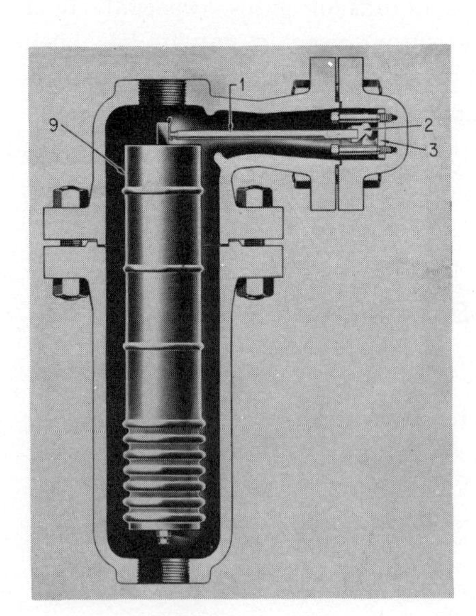

Fig. 13-1 Displacer cage, displacer, and torque-tube assembly. (*Fisher Governor Company*)

tube 4. Torque tube 4 consists of ⅛-in.-diameter rotary shaft 5 welded to female socket 6 inside the tube. Rotary shaft 5 extends through the length of the torque tube, through the outer-tube flange, and into the pilot case which is mounted on the torque-tube housing member.

When the unit is completely assembled, as shown in Fig. 13-2, slotted fitted socket 6 mates with float-rod driver 2. Driver bearing 3, which retains the float-rod driver, is bolted solidly to the torque-tube housing. The other end of the torque tube is held firmly in position at its flanged end by means of retaining flange 7 bolted onto the torque-tube housing.

With no liquid in the float cage, the weight of float 9 (see Fig. 13-1) exerts a downward force on the free end of float rod 1. This causes a turning movement in the torque tube which is equal to the weight of the float multiplied by the length of the float rod. Since the flanged end of the torque tube is clamped to the housing and slotted fitting 6 is mated onto the float-rod driver, which is free to rotate, the torque tube is twisted throughout its length. Suppose that the float cage is filled with water. The float is buoyed up by a force equal to the weight of the liquid it displaced. This action decreases the turning moment being applied to the torque tube, causing it to untwist proportionately. Since rotary shaft 5 is welded to female socket 6, the shaft is rotated as the tube untwists; thus any movement of the float results in rotation of the rotary shaft. Suppose that the float is filled with water and its weight adjusted so that the total weight is exactly equal to the weight of the water which is displaced by it. The torque tube would be in neutral position with no twist in either direction. If the water in the

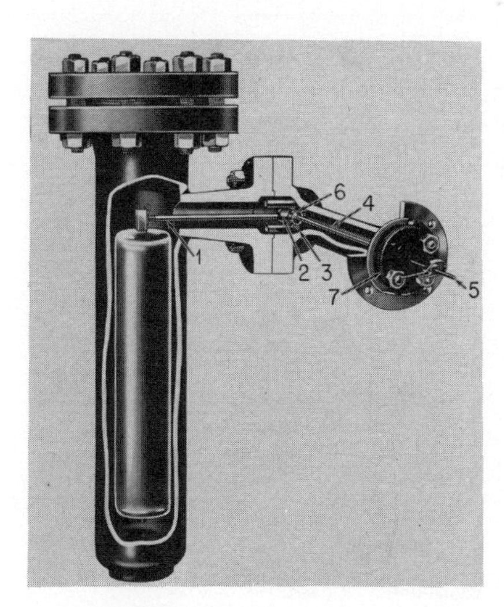

Fig. 13-2 Torque displacer, and cage of Fisher Governor Company Level-Trol. (*Fisher Governor Company*)

float cage is replaced by a liquid whose specific gravity is above that of water (e.g., 1.20), the float is then buoyed up as if it had lost weight. This upward force twists the torque tube, causing the shaft to rotate a certain amount. If the liquid in the float cage is replaced with a liquid lighter than water (e.g., with a specific gravity of 0.80), the float would have less buoyancy and would act on the torque tube as if it had gained weight and would rotate the shaft in the opposite direction.

Thus the shaft responds to every change in the weight of the float, and the apparent weight of the float is proportional to the specific gravity of the liquid which surrounds it. Therefore the position of the shaft is determined by the specific gravity of the liquid in which the float is submerged. Since the specific gravity of all liquids changes with temperature, some means must be used to compensate for it. In process control the float is filled with a liquid that has the same thermal-expansion coefficient and very nearly the same specific gravity as the liquid to be measured. In this manner the temperature of the liquid surrounding the float can be compensated for. Since the specific gravity of liquids decreases with an increase of temperature and expands with a temperature increase, it is obvious that the temperature of the liquid surrounding the float would cause the float length to increase and decrease with temperature changes by expansion and contraction of the corrugated bottom section (see Fig. 13-3).

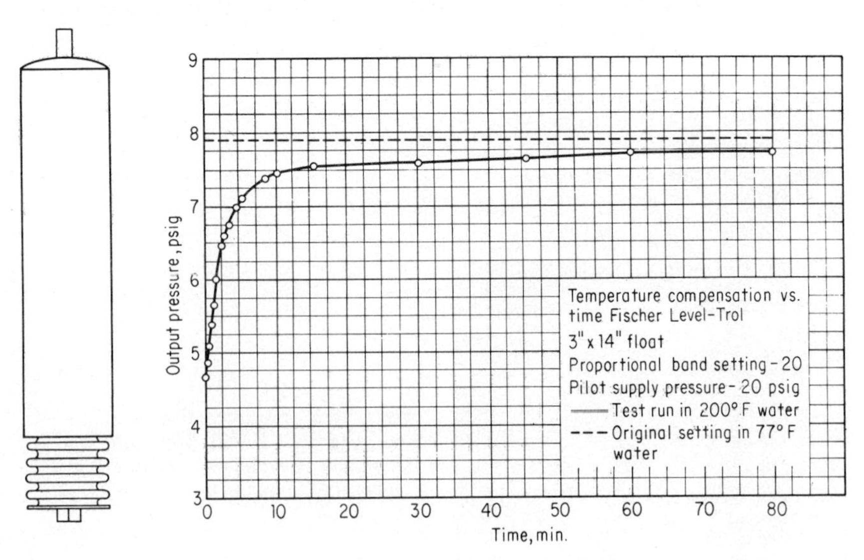

Fig. 13-3 Heat transfer through wall of displacer. (*Fisher Governor Company*)

Suppose the specific gravity of a liquid is being measured. The float fills with some of the same liquid which is being measured, and the temperature of the liquid increases, which causes a decrease in the specific gravity. This causes an apparent increase in weight of the float, but owing to the thermal expansion of the liquid in the float, its length increases the right amount to cause it to displace additional liquid to balance the effect of the specific-gravity change caused by the temperature change. There is a time lag between the temperature of the liquid in the float and that surrounding it. Figure 13-3 shows the rate at which the temperature of the liquid in the float changes and the time required for full compensation. Another advantage of the corrugated section of the float is its ability to change its dimensions with pressure changes, thereby compensating for any change in specific gravity caused by pressure.

It has been shown that the shaft position is dependent upon the specific gravity of the liquid in the float cage. Figure 10-4 shows the method used to convert shaft motion into an air signal, the pressure of which is proportional to the specific gravity of the liquid surrounding the float. The system used to convert the shaft rotary motion into air pressure is a flapper-and-nozzle arrangement. Note that flapper 5 is a modified U-shaped, flat spring which is solidly fastened to the end of shaft 8. As shaft 8 rotates, it moves flapper 5 near or away from nozzle 4, thereby producing a pressure change in Bourdon tube 9, which is transmitted to pilot valve 10, where it is amplified. This amplified pressure becomes the output signal of the instrument, which is indicated by the output gauge. This signal can be recorded on any conventional receiving-pressure recorder which can be calibrated to read in specific gravity. In cases where it is desired to control the specific gravity by blending two process streams, the output signal from the specific-gravity instrument can be transmitted to a receiving controller which in turn operates a control valve located in one of the process streams.

AVAILABLE RANGES: The ranges available in the tubular-float-displacement type of specific-gravity instrument are shown in Fig. 13-4. Note that these ranges are given with a proportional-band setting of 10 percent. Since the range of the instrument is dependent upon the proportional-band setting, other ranges are available when the band is properly set. For instance, an instrument with the proportional band set at 20 percent produces a 1-psi change in output pressure for a change in specific gravity of 0.026. With the band set at 5 percent, a change in specific gravity of 0.006 produces a 1-psi change in output pressure.

In any specific-gravity application, the minimum proportional band or amount of gravity change allowable for a 3- to 15-psi output must be known in order to determine the displacer float required.

Figure 13-4 also shows the output pressure change that would be obtained from a Fisher Governor Level-Trol for a given gravity change for various float volumes at different proportional-band-dial settings. From this chart the proper float size can be selected for any specific-gravity application if the specific gravity of the liquid and the amount of gravity change that can be tolerated are known.

As an example, find the correct float size for the following conditions:

$$\text{specific gravity} = 1.22$$
$$\text{desired range of measurement} = 0.015$$

Refer to the curve in Fig. 13-4, which shows a proportional-band-dial setting of 1.0 for a specific-gravity range of 0.030 and a float volume of about 340 cu in. necessary to obtain a 12-psi (3- to 15-psi) output change from the transmitter. From Table 13-1 note that a float size of 3 by 48 or 4 by 32 in. is satisfactory. The 3- by 48-in. float is close enough to the correct volume, and it could be used by simply changing the proportional-band-dial setting a slight amount. However, the float length might present installation problems, and thus its use under certain conditions is prohibited. The 4- by 32-in. float would be satisfactory from a volume point of view, but since it cannot be used in the standard Fisher Governor Company steel-fabricated cage unit, its manufacturing cost might also prohibit its use.

The next choice in a float size for this specific-gravity application would be a torque tube of light construction (see Table 13-2).

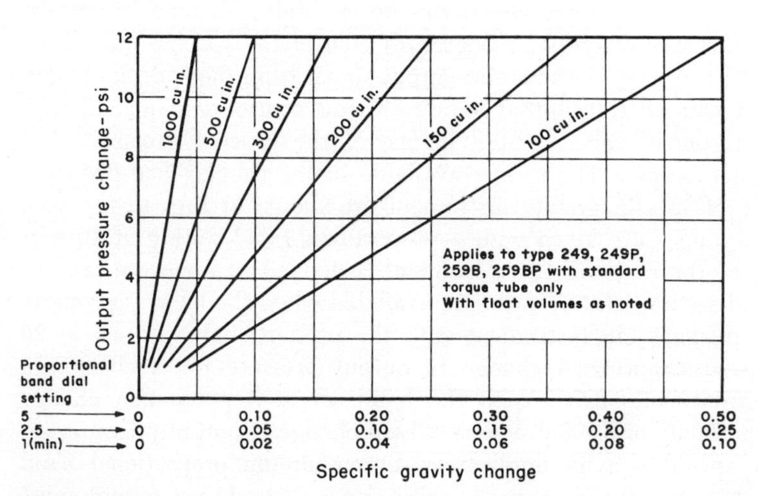

Fig. 13-4 Specific gravity available ranges. (*Fisher Governor Company*)

TABLE 13-1 Volume of Various-size Displacement Floats, Fisher Governor Company

Displacer float size, in.	Volume, cu in.
3 × 14	100
3 × 32	226
3 × 48	339
4 × 14	176
4 × 32	402
4 × 48	603
6 × 14	396
6 × 32	905
6 × 48	1,360

The light torque tube would reduce the required float volume by half, or to 170 cu in. Note that in Table 13-1 a 3- by 32-in. float has a volume greater than the required 170 cu in. Thus the proportional-band-dial setting for the light torque tube and 3- by 32-in. float can be calculated as follows:

$$\text{dial setting} = \frac{(\text{vol. of float})(\text{sp-gr range})}{5^*}$$

Substituting in the formula, we find the dial setting for example to be

$$\frac{226 \times 0.30}{5} = 1.35 \text{ dial setting}$$

The cage-type Level-Trol can be furnished with a piezometer ring if necessary in order to eliminate velocity effects caused by the sample stream flowing through the float cage. The use of a piezometer should

* 5 is used as a divisor when a light torque tube is used for a 3- to 15-psi output for the selected specific-gravity range. If a standard torque tube is used, then 10 is the divisor.

TABLE 13-2 Standard Force of Level-Trols, Fisher Governor Company

Style of torque tube	Displacement force, fb, lb
Standard	0.36
Light	0.18

be practiced on all specific-gravity-transmitter applications where the gravity is measured in a flowing line and where the velocity exceeds 2 fpm.

The piezometer ring allows liquid to flow into the float cage near the center and out through the two equalizing connections. Valves in the two equalizing lines should be used to regulate the flows and to keep the float cage filled. It is also advisable to provide a rotameter or a sight-flow fitting for determining the velocity through the cage.

Pressure limitations. The pressure limits at which a tubular-float-displacement type of gravity meter can be used depend upon the float-cage design if it is an externally mounted unit. Some manufacturers have standard equipment designed to operate at 2,500 psi. As previously stated, the float is completely filled with a liquid and is corrugated so it can expand and contract with pressure changes, which means that the pressure in the float is always equal to the pressure surrounding it. Therefore, the pressure at which the float can operate is unlimited. Other factors rather than the float would determine the pressure limits for safe operation. These limitations are set by the manufacturer of the equipment.

Temperature limitations. The temperature limitations are based on the ASA rating of the flanges of the cage assembly; that is, the pressure rating is based on the temperature at which the flanges operate.

The internal construction, such as the torque tube and float, operates at any temperature that is safe for the cage flanges.

Transmitting, Pneumatic

The pneumatic transmission system is more popular than the electrical with this type of instrument. One of the most important advantages of the pneumatic system over the electrical is its simplicity. The transmitter usually consists of a simple flapper-and-nozzle arrangement with a pilot valve (see Fig. 10-6). The output can be received by a simple pressure-measuring instrument which can be of the indicating or recording type. If a controller is required, the receiving instrument can be equipped with a pneumatic control mechanism. Calibration of the transmitter and the receiver can be checked with an accurate low-pressure test gauge by any person familiar with pressure measurement. However, the pneumatic system is not practical for long-distance transmission where fast response is important for good process control. For such a system 500 ft is a reasonable limit to set.

One item that makes the pneumatic system impractical for some locations is a dry-air supply. In many remote locations, this can be the deciding factor for selecting an electrical transmission system over a pneumatic.

Gamma-ray-type Elements

In-Val-Co, Division of Combustion Engineering, Incorporated

PRINCIPLE OF DESIGN: The gamma-ray-type instrument has proved to be a valuable instrument for measuring the specific gravity of slurries, corrosive liquids, and liquids in high-pressure vessels and pipelines.

The price of the instrument prohibits its use in many cases for measuring the specific gravity of liquids and slurries in open or vented tanks since such measurement can be made with the tubular-float-displacement type or the differential-pressure type, which are much lower in cost.

The gamma-ray instrument finds its greatest application in measuring the specific gravity of liquids in vessels and pipelines operating under pressure from a few hundred to several thousand pounds.

The principle upon which the gamma-ray machine is designed to operate is the measurement of gamma rays emitted by some radioactive material. There are two methods for measuring these rays by industrial-type instruments. One is to intercept the rays with a geiger-counter tube shown in Fig. 13-5 filled with a halogen gas and to use an electrical quench circuit. For normal ranges, the count rate is represented by 0- to 40-mv dc output which is applied to a conventional recording

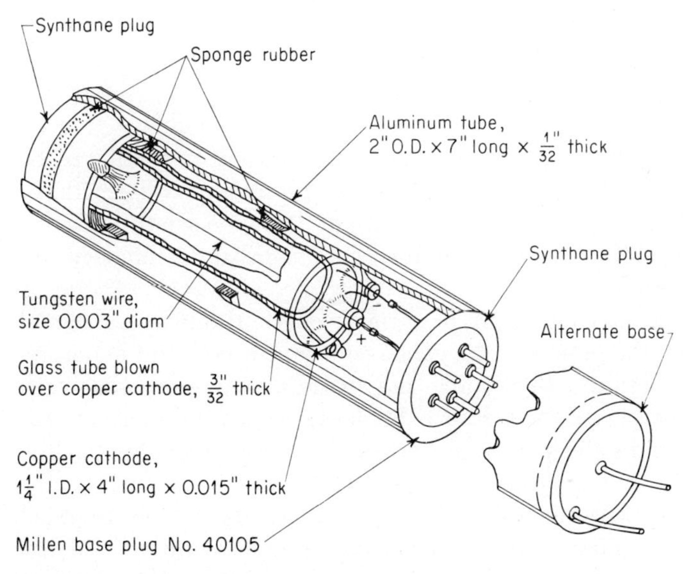

Fig. 13-5 Pictorial drawing, geiger-counter tube. (*In-Val-Co, Division of Combustion Engineering, Inc.*)

potentiometer as its input signal. The instrument can be operated from background to approximately 4,000 counts per sec.

Radiation response of the geiger counter is 80 counts per sec above background for 1 mg of radium at 1-mi distance. By a change of the position-range control, the instrument has a range from background to 3,600 counts per sec, which can be divided into 10 equal increments.

The relationship between absorption of gamma radiation and the path length through the absorbing material is logarithmic. This means that if a certain path length (e.g., 2 in.) absorbs 50 percent, then an additional 2 in. of material will absorb 50 percent of the remainder, or 75 percent absorption in all. The 50 percent absorption thickness of a few key materials given below is useful information for quick estimates of this distance.

5.67 in. for water
2.52 in. for average concrete
0.8 in. for iron
0.547 in. for lead

This holds true for gamma radiation from radium. If other source materials are used, the absorption factor varies, depending upon the activity of the course involved. Since 50 percent absorption distance and the densities vary in a lineal manner, X_1 and X_2 are the half-value distance for X materials having density values D_1 and D_2:

$$\frac{X_1}{X_2} = \frac{D_1}{D_2}$$

With the use of this formula, the half value of any material can be calculated if its density is known. Since the absorption is a logarithmic function of path length, it will plot as a straight line on a semilogarithmic draft paper. Therefore, one value for each material is sufficient for plotting its absorption

It now becomes obvious that if a path length through a material remains constant, the output of the geiger tube will vary with the density of the material between the source and tube, which means that such an instrument can be used to measure the density of certain solids and/or liquids accurately in the path between the gamma-ray source and the geiger tube.

When applying instruments of the gamma-ray type, bear in mind that the radiation intensity varies inversely as the square root of the distance from the source to the counter and directly as the amount of radioactive-source material.

THE OHMART CORPORATION

PRINCIPLE OF DESIGN: The Ohmart Corporation gamma-ray instrument uses a cell which is energized by a cesium source. The cell has an electrical output proportional to the intensity of gamma rays striking it, thereby converting radioactive energy directly into electrical energy.

This type of instrument has two significant features. One is the elimination of high voltage at the point of measurement, which is necessary with the geiger-counter tube. The other is the compensating cell which is used to nullify or "buck out" the current generated by the measuring cell when it is measuring zero level.

In the design of a liquid-level or a specific-gravity-measuring system in high-pressure processes where it is desirable to use a gamma-ray instrument, the required amount of radioactive material must not exceed the amount approved by the U.S. Atomic Energy Commission (AEC) for field installations.

Radiation safety. Most of the radioactive material used in radioactivity gauges is regulated by the AEC which issues a license to users and manufacturers of radioactive materials and inspects sites where radioactive materials are used to determine that the material is being handled properly and safely. The AEC has issued rules and regulations governing the licensing of radioactive materials and radiation safety.

During 1962, the AEC initiated a program to allow the individual states to control radioactive materials. The state regulations are essentially the same as the AEC regulations and, in many instances, have identical wording. Thus, even if the state in which the gauge is used has its own regulations, the AEC regulations are generally applicable.

Gamma-emitting, radioactive materials radiate electromagnetic energy similar to light, except that it is much more penetrating. Radioactive energy can "see" through several inches of steel or similar dense materials. This ability to penetrate dense materials can be used to advantage in the measurement of density, level, and specific gravity of process materials.

Radioactive energy is harmful to the human body only when it is absorbed at an excessive rate. For example, a glowing incandescent lamp cannot be held in the hand without its producing a painful burn. The hand can be held one-quarter of an inch from the lamp for seconds, or inches from the lamp for hours, or several feet from the lamp continuously. Of course, an insulating jacket could be placed around the lamp so that the hand could be in continuous contact with the jacket.

Radioactive energy and radiation are analogous to light energy and radiation with the radioactive source taking the place of the incandescent lamp. Permissible human exposure to a radioactive source is dependent upon: the number of millicuries of radioactive material in the source (similar to wattage rating of an incandescent lamp), distance from the source, and amount of absorber between the sources and body.

The term "milliroentgen per hour" (abbreviated mR/hr) is a measure of the radiation field intensity in air. When radiation is absorbed by the body, the term "rem" or "milliren" (0.001 rem, abbreviated mrem) is used. This distinction is necessary because not all radiation affects the body in the same manner. For gamma radiation the milliren is equal to the milliroentgen.

The AEC limits the amount of radiation which a person could receive to 1.25 rem per calendar quarter. This is an average of about 100 mrem per week.

The 1.25 rem per calendar quarter limitation is a dose at which there is no possibility of injury. However, since the use of gamma radiation is relatively new, the history of injury is not complete. Thus, it is wise to receive as little radiation as possible. To guard against possible overexposure and to maintain a record of personnel routinely exposed to radiation, the AEC requires monitoring of personnel who are apt to receive more than an average of 25 mrem per week or who are exposed to a radiation field greater than 100 mR/hr. When personnel monitoring is required, a record must be kept showing the dose received. When records are kept, and if the employees request it, the employer must furnish a written report of radiation exposure annually and on termination of employment.

In the majority of Ohmart installations, the source is contained in a lead-filled-source holder with an Off and an On position. The holder is designed so that the radiation field is 5 mR/hr (or less) at a distance of 12 in. from the surface of the holder when it is in the Off position. When the source holder is mounted on the pipe or vessel and turned to the On position, the pipe walls, the process material, and the mounting brackets absorb most of the radiation, and again the field intensity is about 5 mR/hr at a distance of 12 in. from the surface of the gauge. Thus, a person would have to be within 12 in. of the gauge for 20 hr per week to receive 100 mrem. A person would have to be within 12 in. of the gauge for 5 hr per week before he would be required to have a personnel-monitoring device such as a film badge or a dosimeter.

Long experience with hundreds of gauges, where the source is contained in a source holder, indicates that the dose received by operators, maintenance personnel, and supervisors is less than an average of 25 mrem per week. Thus, for gauges where the source is contained in a source holder it is usually not necessary to provide any personnel with monitors. Since monitors are not required, obviously no written records must be kept, and employees need not be advised of their radiation exposure.

In some installations, it is impossible to mount the source in a source holder. In these cases the source is usually mounted in a source well. Installation of the source in the well should be done as rapidly as possible. All the necessary equipment should be assembled before the shipping box containing the source is opened. A trial installation using a "dummy" source is recommended. A dummy source can easily be fabricated from steel or brass, using the outline drawing of the source supplied by the Ohmart Corporation. When an unshielded source is installed in a vessel or when it is wipe-tested, the radiation field is usually grater than 100 mR/hr; thus, personnel monitoring in the form of film badges or dosimeters is required. A record of the film badge or dosimeter reading must be kept on a form. Since records must be kept, the employer must furnish to the employee, if he requests it, a record of his radiation exposure annually and on termination of employment.

Additional precautions are required when a gauge is used on a vessel large enough to permit entry of personnel. With the source holder in the open position, or when the source is not removed from the source well, the radiation field intensity inside the vessel can be fairly high. A procedure should be established so that personnel do not enter the vessel until the source holder is in the closed position or the source removed from the source well. The use of padlocks on all man-way and access port covers is acceptable. The key or combination for the locks should be kept by the person responsible for radiation safety.

In some cases, when the vessel or pipe is empty the radiation field intensity on the outside of the pipe or vessel is such that personnel monitoring is required. For installations using source holders, this problem can easily be solved by turning the source holder to the Off position. For installations using sources in source wells where the radiation cannot be turned off, it may be desirable to remove the source temporarily and return it to its lead-shielded shipping-and-storage container.

The best method for determining the radiation field intensity is by measurement with a survey meter. However, the field intensity can be calculated fairly accurately.

Field-intensity calculation. The radiation field intensity can be calculated from:

$$D = \frac{\mathrm{KMC}}{(d)^2} \times 1{,}000$$

D = dose rate, mR/hr
MC = millicurie value of source
d = distance to source, inches
K = constant, 1.3 for radium, 0.6 for Cs-137, 2.0 for Co-60

Suppose that for a certain installation the estimated exposure time to the unshielded source is 10 min at an average body-to-source distance of 20 in. The source is 10 millicuries of Cs-137.

For the above example the dose rate would be

$$\text{dose rate} = \frac{0.6 \times 10}{(20)^2} \times 1{,}000 = {}^{6}\!/\!_{400}\ 1{,}000 = 15\ \text{mR/hr}$$

And the dosage received would be

$$\text{total dose} = 15 \times {}^{10}\!/\!_{60} = 2.5\ \text{mrem}$$

When the radiation field intensity on the outside of the vessel (with the source installed in the source well) is calculated, a set of transmission curves is needed. The graph in Fig. 13-6 shows the percentage transmission for Cs-137 and Co-60 vs. material thickness for a variety of materials. Note that the material-thickness scale is different for lead and steel.

Calculate the radiation field intensity at 12 in. from the surface of a vessel when it is filled with water.

$$\text{total distance} = 12 + 12 = 24\ \text{in.}$$

Dose rate for unshielded source:

$$D = \frac{0.6 \times 10}{(24)^2} \times 1{,}000 = {}^{6}\!/\!_{576} \times 1{,}000 = 10.4\ \text{mR/hr}$$

The percent transmission of gamma radiation through ¼ in. steel (vessel wall) is 0.113 in. Hastelloy (source = well wall) + 12 in. of water = $0.90 \times 0.96 \times 0.335 = 0.289$. The resultant field intensity is $10.4 \times 0.289 = 3.01$ mR/hr. Note that when the vessel is empty the field intensity is $10.4 \times 0.90 \times 0.96 = 8.98$ mR/hr.

When the dosage received by personnel working in the vicinity of the source is estimated, the occupancy must be known. Suppose that a man worked 10 hr per week within 12 in. of the vessel. He would receive a dose of 30.1 mrem per week with the vessel filled with process material and 89.8 mrem per week when the vessel is empty.

For those who desire further information on radiation safety and the handling of radioactive materials, the following publications are recommended: Hine and Brownell, *Radiation Dosimetry*, Academic Press, Inc., New York, New York. These National Bureau of Standards handbooks are available from the Superintendent of Documents, Washington, D.C.: no. 42, "Safe Handling of Radioactive Isotopes," 1962; no. 54, "Protection against Radiations from Radium, Cobalt-60 and Cesium-137," 1962; and no. 59, "Permissible Dose from External Source of Ionizing Radiation," 1962.

The material contained in AEC regulations is too detailed to include here. It deals with the AEC and contractor agreements; however, the user is provided with the necessary material by the vendor at the time of the delivery of the gamma-ray sources.

Figure 13-6 shows the thickness and absorption of gamma rays by various materials for a 1-mg radioactive source.

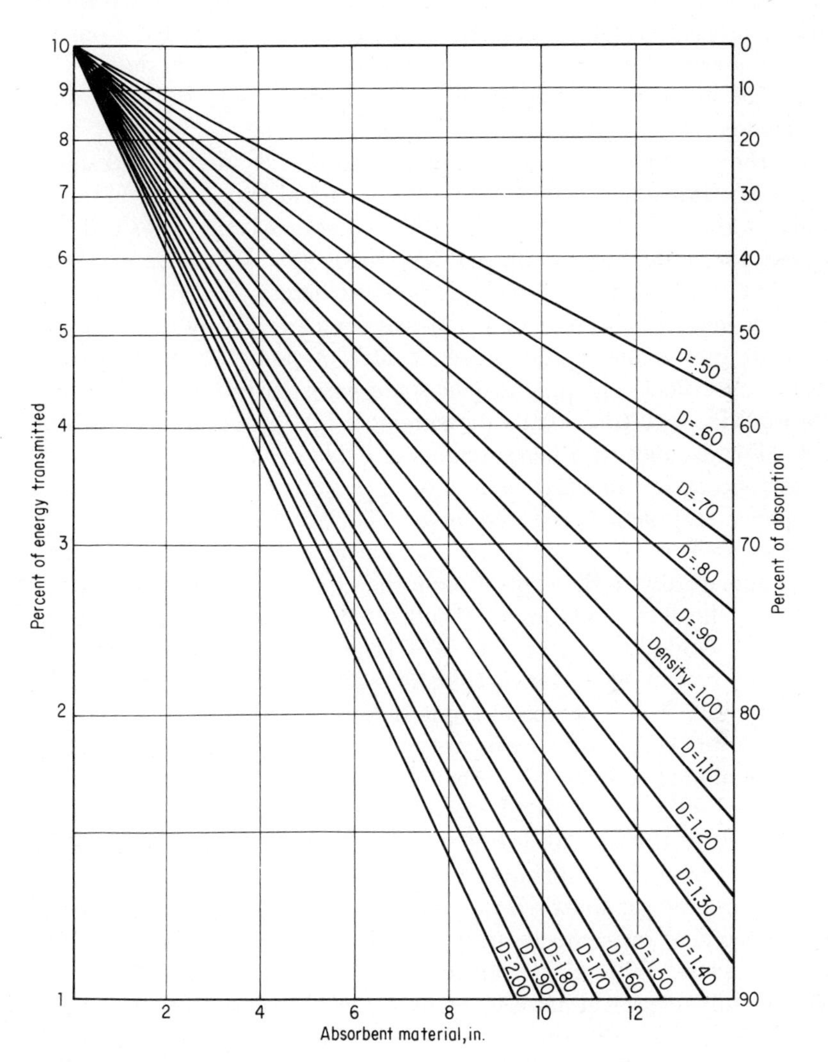

Fig. 13-6 Gamma-ray absorption by various density of materials.

In the majority of processes liquid-level measurement can be made by a gamma-ray instrument without the radioactive source within the vessel or tank in which the level is to be measured being installed. Therefore none of the components are exposed to the temperature of the process material. Good engineering practice requires that no instrument-measuring circuit be exposed to ambient temperatures above 150°F even though it can operate at temperatures well above this. Servicing and maintenance are simplified if the temperature is kept below 150°F.

For installations where it is necessary to install the radioactive source within a process vessel or tank, the source must be protected from the process material. If the source is in the form of a cobalt slug or one of the radium salts, a holder similar to a thermocouple well will serve the purpose.

Since the output from a gamma-ray-pickup unit is an electrical signal, the primary instrument must be of the electrical type. The electrical circuits of the geiger-tube-type and the direct-conversion-type cells are designed for a dc voltage output less than 100 mv to the receiving instrument, which can be a simple dc potentiometer.

When a level or specific-gravity measurement made by a gamma-ray instrument into a pneumatic process-control system is incorporated, it is more convenient and practical in some cases to use an electropneumatic transducer to convert the measurement into a standard 3- to 15-psi signal. For instance, if a measurement is made in a nonhazardous area and it is necessary to cascade it into a pneumatic control system for safe operating practice and the process-control room is located in a hazardous area, it would be more economical to transmit the measurement pneumatically to the control room than to install an explosion-proof receiver in the hazardous area. Another case where electropneumatic transducers would be practical is in a large, complex, pneumatic process-control system where an automatic scanning-and-logging device is used to operate from pneumatic input signals. Transducers would permit integration of all measurements made by gamma-ray instruments into the scanning system. In other cases, for the sake of flexibility, it might be desirable to keep all controlling instruments of the pneumatic type, thereby justifying the use of the electropneumatic transducers for measurement made by gamma-ray instruments.

Because the output signals from gamma-ray-pickup units are electrical and the receiving instruments are potentiometers, transmission over distances of several miles is practical. The potentiometers may also be connected into electrical-impulse scanning-and-logging devices without their affecting the receiving instrument if the input resistance of the scanning device is 10,000 ohms or greater. Gamma-ray-pickup units may also be cascaded into process-control systems where electronic receiving control instruments are used, provided the input resistance exceeds the 10,000-ohm value.

The gamma-ray level-measuring instrument definitely has a place in process-control systems where the level must be measured of such materials as slurries, highly corrosive and toxic liquids, granular catalysts, and other dry materials and sludge formations in vessels or tanks and where it is not practical to have the measuring equipment come in contact with the process material. Another important application is the measurement of level in process vessels operating at pressures of several thousand pounds per sq in. Besides enabling measurement of level of various materials in vessels and tanks, the gamma-ray instrument is also valuable in detecting foaming in chemical reactors and liquid carry-over with vapors from quench towers, vaporizers, and similar processing equipment.

A motor-driven, gamma-ray level-measuring instrument has been developed by the Ohmart Corporation for applications where it is desired to measure liquid level accurately over wide ranges, and because of physical conditions, mechanical means are impractical. Shown schematically in Fig. 13-7 is the principle upon which the Ohmart model MDL gauge operates. Note that a source of gamma rays and a measur-

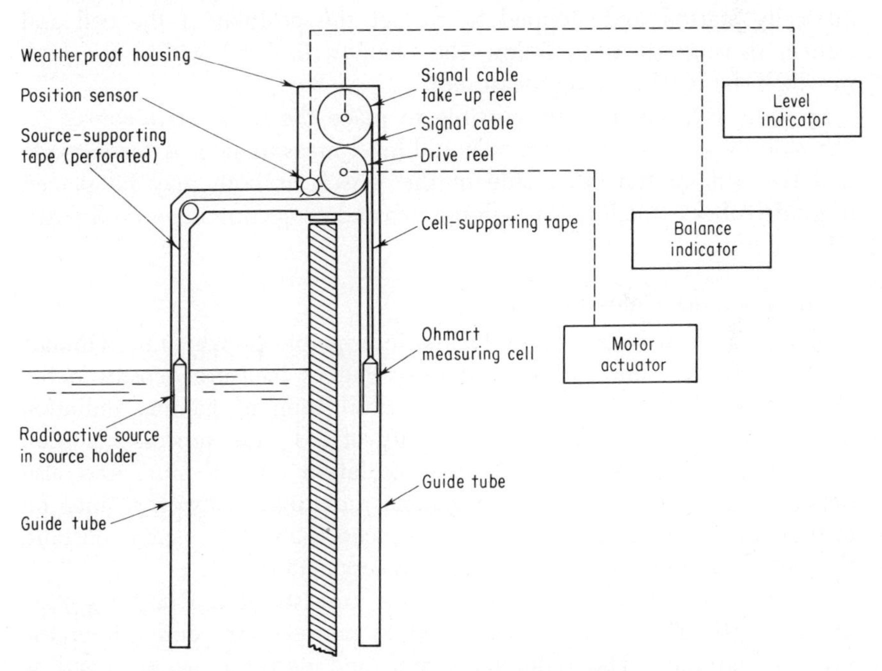

Fig. 13-7 Schematic view of liquid-level measurement by a motor driven source and pickup. (*The Ohmart Corporation*)

ing cell are mounted on steel tapes that wrap around a motor-driven drum. One tape is equipped with perforations that engage the position-sensor sprocket. Thus as the motor-driven drum rotates, the source and measuring cell are automatically moved up and down together, and the position sensor is caused to rotate. The position sensor operates (through a gear-reduction mechanism) two potentiometers which produce output voltages that are a function of the position of the source and cell. One potentiometer indicates the full range in feet; the second indicates the inches within that range. The weather-proof housing also contains a spring-loaded, take-up reel so that the signal cable is maintained at a given mechanical stress as the measuring cell is moved up and down.

A zero-suppression circuit is adjusted so that the balance-indicating meter indicates zero when the level is at midpoint of the measuring cell and source. As the level moves up, the balance indicator points to off balance; thus by means of the motor actuator, the source and cell are moved up correspondingly until the unit reading returns to balance. The new level is then read on the voltmeters constituting the level indicator.

When automatic following is desired, a relay is connected between the balance indicator and the motor actuator so that the motor is automatically started and stopped to correct the position of the cell and source, causing them to follow the changes in level within the tank in which the level is being measured.

In some cases it may be possible to place the radioactive source on one side of a vessel in which a liquid-level measurement is to be made and the cell on the other side of the vessel, or both may be placed in guide tubes extending through nozzles and projecting them both down into the liquid.

The Ohmart Corporation

PRINCIPLE OF MEASUREMENT: The basic principle by which the Ohmart model CP-, CL-, and CS-series density gauges measure density is the absorption of gamma radiation. The absorption of gamma radiation depends upon the thickness and density of: (1) the process material, (2) the process pipe, and (3) the insulation on the pipe, and also depends upon the energy of the gamma radiation. However, since all of these factors (except density of the process material) remain constant, the gauge measures the density of the process material.

The Ohmart model CS density gauge consists of a straight section of pipe with 90° elbows at each end to permit entry and exit of the process material. The radiation source and detector are arranged so that the radiation is directed along the axis of the straight section of

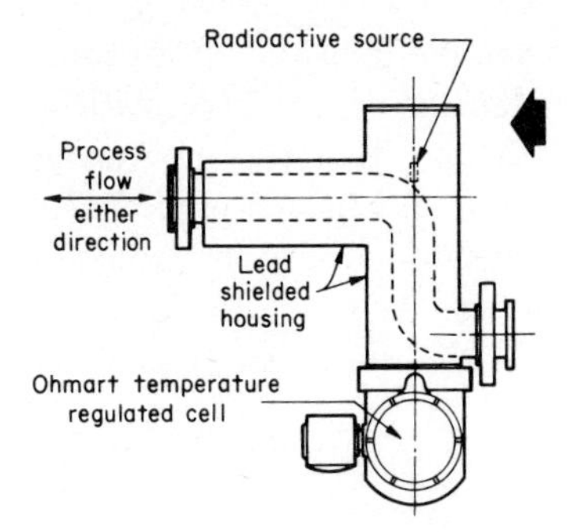

Fig. 13-8 Density gauge for small diameter pipe. (*The Ohmart Corporation*)

pipe. Such an installation is shown in Fig. 13-8. The radiation is provided by a Cs-137 gamma source. The detector is an Ohmart temperature-regulated ionization cell.

The Ohmart model CP- and CL-series density gauges are mounted so that the radiation passes through the diameter of the pipe as shown in Fig. 13-9. Mounting brackets are provided so that the gauge can be clamped onto an existing pipeline. The Ohmart model CP- and CL-series gauges also use a Cl-137 gamma source and an Ohmart temperature-regulated ionization cell.

The Ohmart ionization cell, when exposed to gamma radiation, generates a current which can be measured by the Ohmart model WA or the Ohmart model DPS-1 amplifier.

To provide a null system of measurement, current suppression is used. Suppression provides a current of opposite polarity to that of the measuring cell. The source of this suppression is either a compensating ionization cell or a stable voltage supply. Both of these configurations are shown in block diagram in Figs. 13-10 and 13-11. Note that the amplifier measures the algebraic sum of the currents from the measuring cell and the suppression source.

The suppression current is adjusted to nullify the current generated

Fig. 13-9 Density gauge mounted on pipe line. (*The Ohmart Corporation*)

by the measuring cell for any predetermined sample density. For example, suppose it is desired to measure the density of a slurry with a variation in specific gravity of 1.0 to 1.2. Assume that the measuring cell generates 10 units of current for a gravity of 1.0 and eight units of current for a gravity of 1.2. The suppression current is adjusted to balance the current of 10 units. When the gravity is 1.2, the measuring cell is generating only eight units while the suppression is fixed and is producing 10 units. Thus, a difference in current of two units results for a gravity of 1.2. The amplifier sensitivity is adjusted to produce full-scale deflection for the two units of current. The result is that the meter indicates zero for a gravity of 1.0 and full scale for a gravity of 1.2.

The current suppression control or adjustment screw of a compensating cell is located on the front panel of the instrument. Clockwise rotation of the screw provides increasing current, and counterclockwise rotation, decreasing current. The meter on the amplifier is polarized to read upscale with an increase in suppression current or a decrease in current from the measuring cell.

DESCRIPTION OF COMPONENTS: The measuring assemblies of the Ohmart model CP or CL gauge are shown in Figs. 13-10 and 13-11. The source of radiation, Cs-137, is placed on one side of the pipe which carries the process material to be measured. The detector, an ionization cell, is located directly opposite the source on the other side of the pipe.

The ionization cell generates a current which is measured by either the Ohmart model WA or the model DPS-1 amplifier. The output of the amplifier may then be fed to a controller or recorder for continuous monitoring.

GENERAL NOMENCLATURE: The CP series is for pipes 2 through 8 in.; the CL series is for pipes 10 through 20 in. There is no essential difference in the gauge components for these two series except that in the CL series, the model HM-8 source holder is provided with additional collimation. The complete model number designations which provide

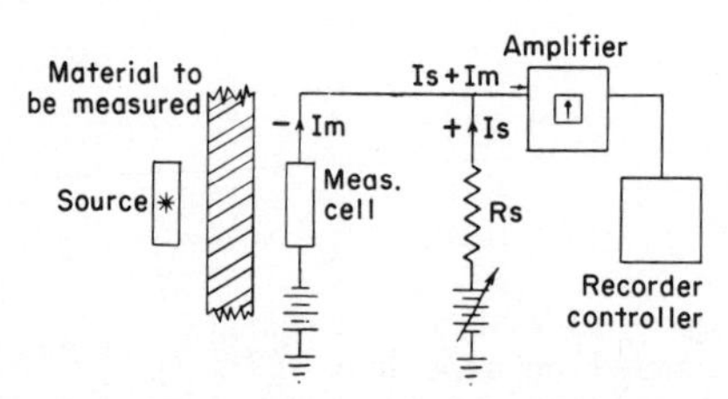

Fig. 13-10 Density gauge using electrical suppression. (*The Ohmart Corporation*)

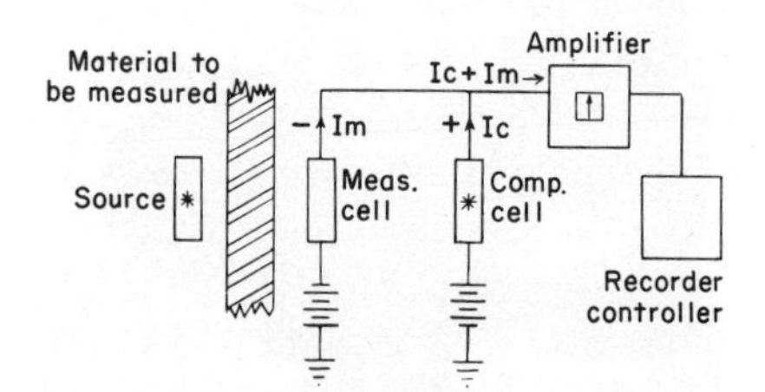

Fig. 13-11 Density gauge using compensating cell suppressing. (*The Ohmart Corporation*)

information as to the pipe size and electronic amplifier are as follows: CP-0-WC, CP-0-WE, CP-0-D, CL-0-WC, CL-0-WE, CL-0-D.

The zero represents the pipe size; e.g., 10 for a 10-in. pipe. This may be any number from 2 through 20. WC represents the Ohmart model WA amplifier with a compensating ionization cell providing the suppression. WE represents the model WA amplifier with electrical suppression. D represents the Ohmart model DPS-1 amplifier which always uses electrical suppression.

SOURCE AND SOURCE HOLDER: The radioactive source used in these density gauges is cesium-137, a gamma emitter with an energy of 0.66 Mev and a half-life of 33 years. The radioactive material is contained in a small metal capsule which is sealed to prevent the possibility of leakage of radioactive material and consequent contamination of the gauge and/or area in which it is installed.

The radioactive source is mounted in the model HM-8 source holder which provides shielding of gamma radiation for the safety of the personnel using the gauge. Shielding is such that personnel working in the area where the gauge is installed are not exposed to any risk from overexposure of gamma radiation.

The red handle on the model HM-8 source holder operates a rotating shutter which has two positions—On and Off—and the means for padlocking the shutter in the Off position during shipment and installation.

A yellow and magenta source-data tag is affixed to the model HM-8 to show the type and quantity of radioactive material. In the event of abandonment of the equipment, the model HM-8 source holder must be returned intact, with the radioactive material, to the Ohmart Corporation for proper disposal. Malfunction of the source holder should be reported to the Ohmart Corporation and returned for repair, after receipt of complete handling and shipping instructions from the Ohmart Corporation.

OHMART IONIZATION CELLS: The ionization of cells used on the model CP- and CL-series gauges are the model RTR and model CC cells which are specially designed to measure the "hard" or high-energy radiation from Cs-137. These cells use a very low polarizing voltage across the electrodes to collect the ions produced by radiation striking the cell. These cells inherently possess guarded electrode design, low-leakage insulators, and excellent stability and reliability.

The measuring cell, the model RTR, has a negative polarity; i.e., the current from the measuring cell is negative with respect to amplifier chassis. Since current suppression is used to provide a null system of measurement, the suppression current is positive with respect to the amplifier chassis. If the current suppression is provided by an Ohmart model CC compensating cell, the polarity of that cell is positive. Although the temperature coefficient of a cell is small, a heater jacket thermostatically controlled by an SCR-powered heater supply is provided to maintain the temperature of the cells at 140 ±0.1°F., regardless of external-temperature fluctuations. Figure 13-12 is an illustration of the Ohmart model RTR cell.

OHMART AMPLIFIERS: *Ohmart model WA amplifier.* As stated previously, two different types of amplifiers can be used with this gauge series. The model WA is a feedback amplifier which uses a vibrating capacitor to convert the input dc signal to ac for amplification. After amplification, the ac signal is reconverted to dc. This design offers excellent zero stability because all amplification is done in the ac mode. Both electrical-zero and compensating-cell suppression is available with this amplifier.

The output of the amplifier is 0 to —1 volt dc. It also provides 0 to —50 mv for a recorder and milliampere terminal for 0 to 1 ma into a 1,000-ohm load.

Ohmart model DPS-1 amplifier. The Ohmart model DPS-1 is a dc feedback amplifier which uses carefully aged 5886 tubes for low zero drift. The preamplifier may be placed at, or remote from, the point of measurement. Electrical suppression is always used with this amplifier. The output is 0 to —10 volts dc. It also provides 0 to —50 mv for a recorder.

Cable. The output signal from the Ohmart detectors is about 10^{-10} amp. The coaxial cable, used in the gauges with the Ohmart model WA amplifier to transfer current from the detector cell to the amplifier, is treated with colloidal graphite to reduce electrical noise generated by cable movement and vibration.

Location of controls. Most of the controls are located on the side panel of the amplifier. Some controls are located on circuit boards or on front of the instrument panel.

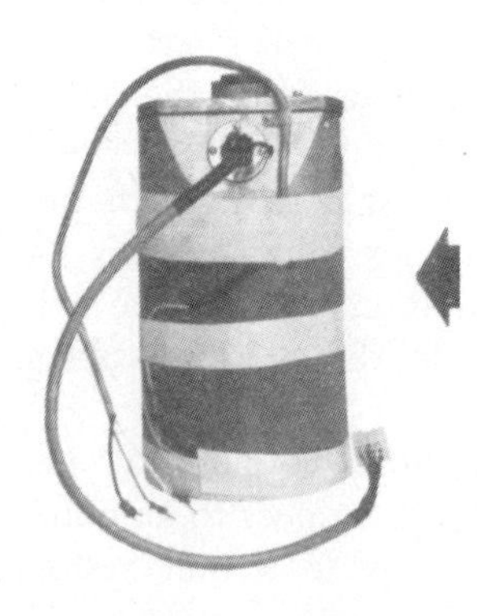

Fig. 13-12 Preamplifier and RTR cell. (*The Ohmart Corporation*)

Equivalent absorbers. Equivalent absorbers are flat sheets of lead, the thickness of which can be adjusted to produce an absorption of radiation equivalent to the process material in the pipe. They are made in the factory and trimmed to the exact thickness in the field. Calibration rechecks can be made with equivalent absorbers simply by emptying the pipe.

Transmission curve and dial settings for predetermined calibration. A transmission curve, with the suppression dial settings, is inserted for use in those gauges where the predetermined calibration technique is employed. This curve is based on data taken with an identical gauge and simulated process material.

The choice of the method for calibration is dictated largely by the application and the user's preference. However, predetermined calibration is not to be used with those gauges which use compensating cell suppression.

Maintenance for the Ohmart system is not included here but may be obtained from the Ohmart Corporation.

GAS SPECIFIC GRAVITY

Float-type Elements

BECKMAN INSTRUMENTS, INCORPORATED*

PRINCIPLE OF DESIGN: Shown schematically in Fig. 13-13 is a model 3A, continuous, gas-density balance designed by Beckman Instrument, Incorporated. Inside the measuring cell a horizontal quartz fiber supports a light dumbbell. One ball of the dumbbell is punctured, making it independent of buoyancy effects; the other ball rises or dips as the density of the gas varies.

Near one of the dumbbell balls, two electrodes held at fixed potentials establish an electrostatic field about the ball. The dumbbell is coated with rhodium to make it electrically conductive. A varying potential applied to the dumbbell subjects it to an electrostatic force within the

* Carroll, *Industrial Process Measuring Instruments,* pp. 238–239.

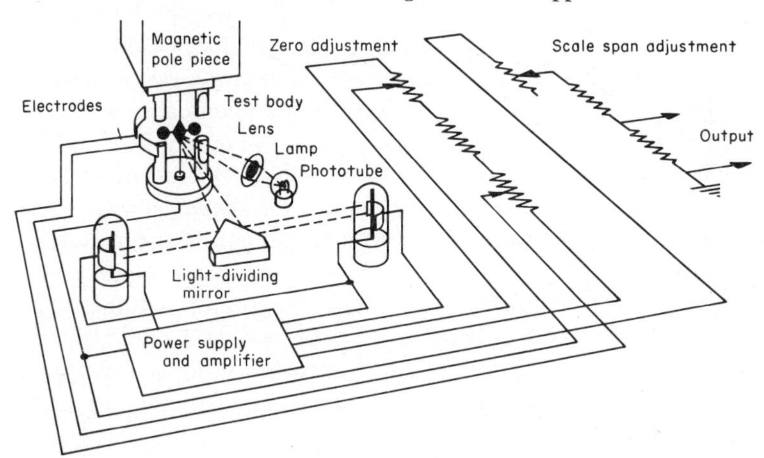

Fig. 13-13 Schematic circuit for specific-gravity recorder. (*Beckman Instruments, Incorporated*)

field. Any change in gas density will exert a force on the dumbbell, causing it either to rise or dip. This force can be counteracted by the application of enough balancing potential to the dumbbell to maintain it in a stationary or null position. The balancing potential output is proportional to the force exerted on the dumbbell by the gas. That force, in turn, is directly proportional to the density of the sample.

The schematic shown in Fig. 13-13 illustrates the method by which the balancing potential is obtained and measured. A light beam reflected from one mirror attached to the dumbbell is divided by another mirror between two phototubes. When the dumbbell rotates, one phototube receives more light than the other, the difference causing a change in their resistance. This change provides a signal which, when amplified, becomes the balancing potential applied to the dumbbell for maintaining its null position. Measuring and indicating this balancing potential with a standard recorder provide linear readings which are proportional to the density of the sample gas.

Response of the instrument is such that 95 percent of any reading within range of the unit is obtained in less than 1 min.

Range spans are available from a minimum of 0.05 specific-gravity units for full scale up to 5.00 relative to air. Multiple ranges are also available in various combinations.

The specific gravity of gases normally considered corrosive can be measured with the Beckman unit by obtaining a balance cell fabricated from a special, noncorrosive material.

Chandler Engineering Company*

PRINCIPLE OF DESIGN: Shown schematically in Fig. 13-14 is a gas specific-gravity-measuring instrument of the Chandler Engineering Company design. The Ac-Me† gravitometer is designed for continuously measuring and recording the specific gravity of natural gas and other noncorrosive gases and can also be equipped with a pneumatic or electrical transmitting unit or controller. It operates on the principle of weighing a volume of gas in comparison with an equal volume of air.

Major components of the recording gravitometer shown in Fig. 13-15 are float, float-balance bar, calibrating weights, pen-arm linkage system, oil seal, and recorder.

The float is cylindrical with a cone-shaped top and bottom. The float-balance bar supports the float, calibrating rod and weights, and compensator mercury-filled U tube and is connected to the recording-pen linkage system. The bar operates on steel knife edges through an angle of 5° for full-scale deflection of the recording pen.

* *Ibid.*, pp. 233–236.

† Registered trade name, Chandler Engineering Company.

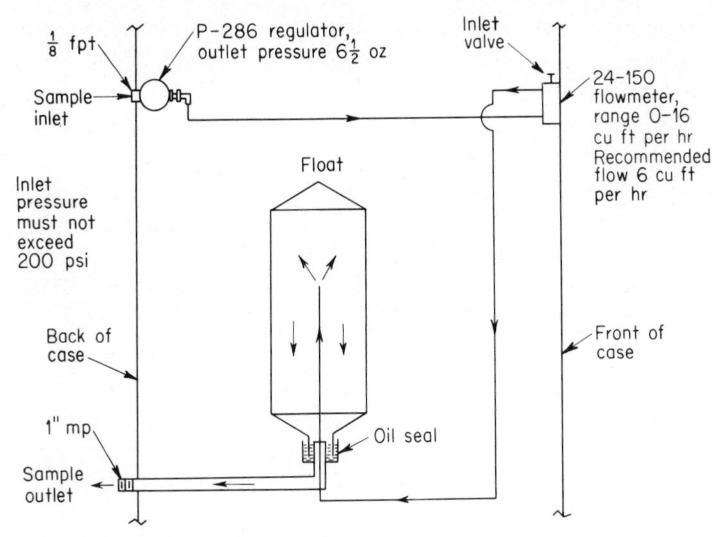

Fig. 13-14 Schematic of gas specific-gravity-measuring instrument. (*Chandler Engineering Company*)

The bottom end of the float extends into an oil-seal cup through which the sample and exhaust tubes pass. A small flow of sample gas is flow-controlled and enters the float through one of the tubes. Baffle plates distribute the incoming gas throughout the float before the sample is discharged into the atmosphere.

The float calibrating weights are adjusted until the vertical position of the float is such that it produces a reading on the chart corresponding to the specific gravity of the gas within the float. Thus the vertical position of the float depends upon the density of the gas which supports a portion of the weight of the float. Any variation in density of the sample gas within the float produces a corresponding vertical movement of the float which is recorded as specific gravity of the sample gas.

So that changing atmospheric conditions can be corrected, an automatic compensator is included in the measuring system. By means of the compensator, the specific gravity as it appears on the recording chart is corrected for atmospheric temperature and barometric pressure.

Compensation is made by means of a mercury-filled U tube, one leg of which is mounted on the instrument case and the other suspended on the balance bar. The former is open to atmospheric pressure, while the latter is connected to a large-volume, temperature-sensing chamber. Changes of pressure and temperature cause the mercury to move in the U tube and change the center of gravity of the entire balanced system in the direction and amount required to correct the chart reading to standard conditions of 760 mm of Hg atmospheric pressure at 60°F.

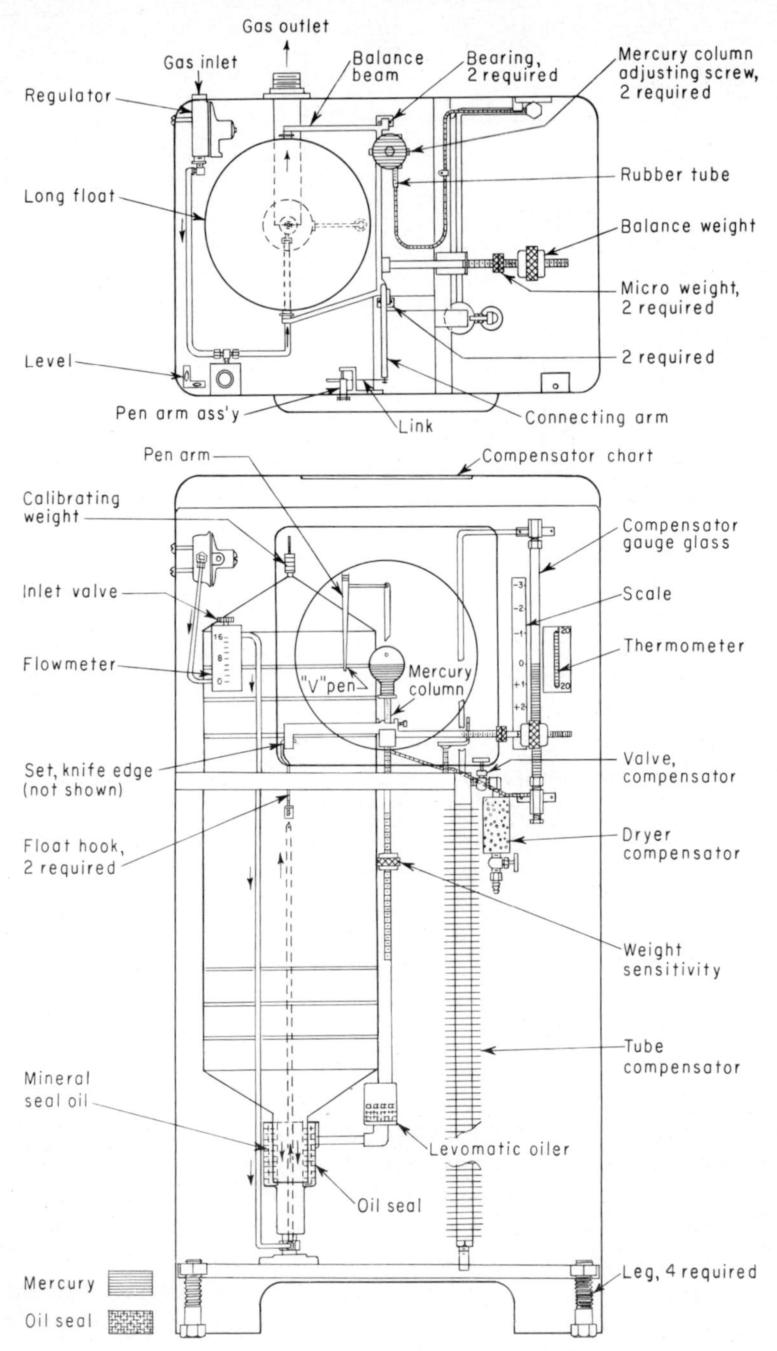

Fig. 13-15 Components of specific-gravity recorder. (*Chandler Engineering Company*)

Chandler Engineering Company*

Principle of design: The gas specific-gravity balance shown in Fig. 13-16 is of the Chandler Engineering Company design and is used to make periodic measurement of the specific gravity of natural gas and other noncorrosive and nontoxic gases. The instrument operates to measure the specific gravity of a gas as defined by the Bureau of Standards: "The ratio of the weight of an equal volume of dry air, free from carbon dioxide, measured at the same temperature and pressure."

Major components of the Ac-Me specific-gravity balance shown in Fig. 13-16 are ball float, float beam, beam suspension, scale, and container. The float beam and scale are suspended by means of four phosphor-bronze suspension springs, ⅛ in. wide by 0.0015 in. thick. There are two suspension springs on each side of the balance beam which are crossed at the top in such a manner that the balance beam has free movement about the point of oscillation but prevents unnecessary lateral motion of the beam. One end of the balance beam is equipped with a millimeter scale which is used in conjunction with a hairline indicator to show when the beam is in balance. The beam is also provided with weights to adjust its sensitivity and balance point.

* Carroll, *Industrial Process Measuring Instruments,* pp. 236–237.

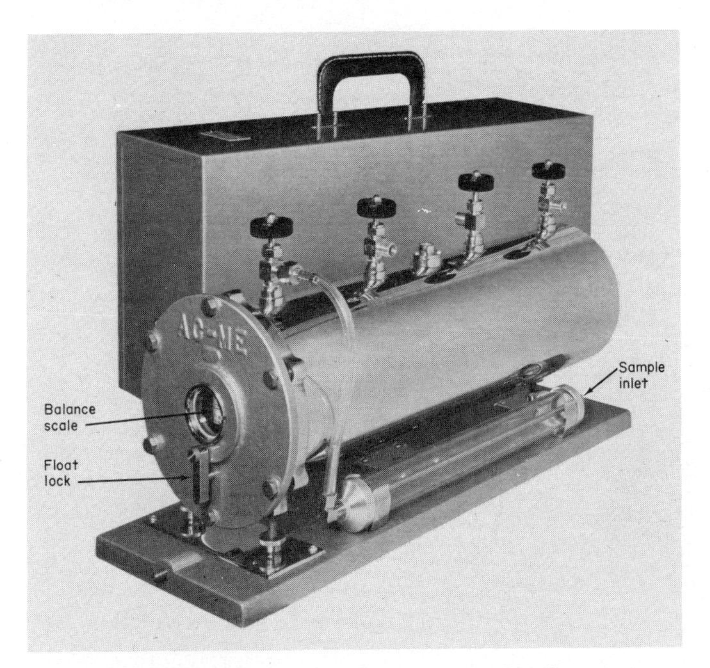

Fig. 13-16 Gas specific-gravity balance. (*Chandler Engineering Company*)

The balance mechanism is contained in a 6-in. brass tube, the front of which is equipped with a glass window through which the beam scale is viewed when a specific-gravity measurement is being made.

The Ac-Me specific-gravity balance operates on the following law of physics: According to Boyle's law, the density of a gas is proportional to its pressure, and the buoyant force exerted upon a body suspended in a gas is proportional to the density of a gas and therefore to its pressure. Therefore, if the buoyant force exerted upon a body is made the same when suspended successively in two gases, the densities of the gases must be the same at these pressures; or the densities of the two gases at normal pressures are in inverse ratio to the pressures when of equal buoyant force.

Potter Aeronautical Corporation*

PRINCIPLE OF DESIGN: Two densimeters have now been added to the Pottermeter† line of basic transducers. Both use the same principle of a resistance change of a potentiometer. The output of the model 27A shown in Fig. 13-17 has an analog output which is in the form of a resistance change of a potentiometer. The output of the model 27D is in the form of a pulse, and therefore it is referred to as a digital densimeter.

* *Ibid.*, pp. 211–213.

† Registered trade name, Potter Aeronautical Corporation.

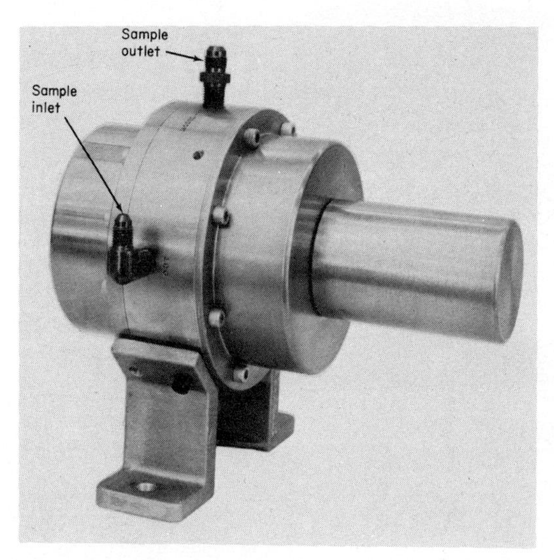

Fig. 13-17 Densimeter of the analog type. (*Potter Aeronautical Corporation*)

Both types are designed to measure continuously the specific gravity of fluid samples under conditions of high pressures and temperatures at a high sampling rate. The range and span of the transducers can be adjusted to meet a wide variety of fluids.

The primary components of the Potter series 27A densimeters are the housing, pendulum, and float assemblies. The float is a submerged, pivoted balance which is positioned by the buoyant force of the liquid. An angular position of the float exists for each fluid density. This angular position is transmitted through the housing by means of a pulsed magnetic circuit. The magnetic circuit is a closed-loop circuit with small air gaps, thereby making it highly efficient. The circuit consists of a permanent magnet and an iron yoke, which is the magnetic follower. The circuit is designed so that the forces of attraction between magnet and yoke are at right angles to the float pivots, thereby eliminating thrust friction on the pivot bearings of the float and output transducer. Surrounding the yoke is a 110-volt 60-Hz coil which serves to vibrate the yoke and float assemblies. This pulsed coupling allows true positioning of the float-and-follower assemblies by eliminating friction from the system.

A balance contains three floats mounted on a common shaft; one float is fixed, and the other two are adjustable. One float is for the span adjustment; the other adjusts the bottom of the span for zero-angle deflection of the float. The floats are constructed of materials that have such an extremely low coefficient of expansion that a change in temperature has little effect on changing the float volumes. A change in float volumes caused by a 100°C change in temperature affects the measurement by 0.0001 part of gravity. Piping connection for bringing the fluid into and out of the float chamber is in the housing. This means that the float can be serviced or removed by simply having the float-chamber cover removed. The fluid is piped into the back of the float chamber between a baffling cylinder and the cover. From the rear, it then passes over and through the float mounting plate to the front of the chamber and out through the housing. This design ensures a continuous change of fluid in the sampling chamber, prevents fluid stagnation, and reduces the effect of turbulence on the float. Since the floats are machined out of solid stock, pressure does not affect accuracy, and the pressure limit is dependent only upon the thickness of material used for that part of the housing through which the magnetic circuit operates. Since this is essentially a small-diameter cylinder, a thin wall at this point can withstand high pressures. To strengthen this section of the housing further, only enough material is removed to allow the magnetic follower to travel through its angular span. The material that is not removed serves as a stop for the follower and a mount for the vibrator coil. The housing and cover can be made as heavy as necessary to meet any practical pressure requirements. On the opposite side of the housing from the float chamber are located the electrical components and pendulum assembly. The function of the pendulum is to reference a permanent magnet and the potentiometer with respect to the earth; since the balancing float also assumes a position with respect to earth, this eliminates the need of accurately mounting the densimeter.

Provision must be made by the customer to maintain a flow of the fluid to be sampled through the densimeter. Maximum flow rate through the densimeter should not exceed ½ gpm. This flow can be accomplished by means of a restriction in the main flow line, an orifice plate, or a suitably chosen pump. The flow through the densimeter should be as high as possible, within the prescribed limit, to assure a continuous, true fluid sample. The fluid lines to the densimeter should be as short as possible and insulated, if necessary, to assure sampling of fluid at pipe temperatures.

It is recommended that a 50-micron filter be employed in the input line to the densimeter (Cuno, commercial "full flow," or equivalent) if an adequate filter is not already applied to the main line ahead of the densimeter and/or flowmeter. In those applications where the densimeter is employed in conjunction with a Pottermeter for mass-flow measurement, the takeoff to the densimeter should be as close to the Pottermeter as feasible.

The Potter densimeter is available for operating over a temperature range from —100 to +500°F at pressures up to 1,000 psig. Densimeters are available for operating at pressures greater than 1,000 psig but are considered optional equipment.

Index